Fakultät Bauingenieurwesen Institut für Wasserbau und Technische Hydromechanik

Dresdner Wasserbauliche Mitteilungen
Heft 59

Roberto Tatis-Muvdi

A CONTRIBUTION TO THE HYDROMORPHOLOGICAL ASSESSMENT OF RUNNING WATERS BASED ON HABITAT DYNAMICS

Der Titel und der Inhalt des Heftes entsprechen der zur Erlangung des akademischen Grades Doktor-Ingenieur (Dr.-Ing.) an der Fakultät Bauingenieurwesen der Technischen Universität Dresden im März 2016 eingereichten und genehmigten Dissertation von M.Sc. Roberto Tatis-Muvdi.

Verteidigung: 30.05.2016

Vorsitzender der Promotionskommission:
Univ.-Prof. Dr.-Ing. habil. Ivo Herle, Technische Universität Dresden

Gutachter:
Univ.-Prof. Dr.-Ing. J. Stamm, Technische Universität Dresden
Apl. Prof. Dr. Jochen Koop, Universität Koblenz-Landau

Dresden, Januar 2017

Bibliografische Informationen

Die Deutsche Bibliothek verzeichnet diese Publikation in der Deutschen Nationalbibliographie. Die bibliografischen Daten sind im Internet über http://dnb.ddb.de abrufbar.

A contribution to the hydromorphological assessment of running waters based on habitat dynamics
(Ein Beitrag zur hydromorphologischen Bewertung von Fließgewässern anhand der Habitatdynamik)

Technische Universität Dresden, Fakultät Bauingenieurwesen,
Institut für Wasserbau und Technische Hydromechanik.

Dresden: Institut für Wasserbau und Technische Hydromechanik, 2017
(Dresdner Wasserbauliche Mitteilungen; H. 59)
Zugl.: Dresden, Techn. Univ., 2017
ISBN 978-3-86780-512-4

Herausgegeben im Auftrag des Rektors der Technischen Universität Dresden von

Univ.-Prof. Dr.-Ing. J. Stamm
Univ.-Prof. Dr.-Ing. habil. K.-U. Graw

Technische Universität Dresden
Fakultät Bauingenieurwesen
Institut für Wasserbau und Technische Hydromechanik
01062 Dresden

Tel.:	+49 351 463 34397
Fax:	+49 351 463 37120
Email:	mail@iwd.tu-dresden.de
WWW:	http://iwd.tu-dresden.de

Redaktion: Roberto Tatis-Muvdi

Redaktionsschluss: 12.01.2017

Verlag: Selbstverlag der Technischen Universität Dresden

ISSN 0949-5061
ISBN 978-3-86780-512-4

No one's free until everyone's free.
- Fannie Lou Hamer, U.S. civil rights activist.

FOREWORD

Humans have always been inextricably linked to water. In particular, rivers have provided concentrations of resources that allowed our species to develop agriculture, focus on things other than mere survival and thus advance art, science and engineering. Ancient Egypt, the Indus Valley civilization and Mesopotamia are examples of this.

However, our use of the natural resources of river landscapes, and of the biosphere in general, has led to species extinction rates up to 100 times higher than the background extinction rate found in nature. Rates of biodiversity loss of this magnitude have occurred only five times in the history of the biosphere; the most famous of these was caused by the asteroid that killed off the dinosaurs 65 million years ago.

Thus, humans have been responsible for the sixth mass extinction on this planet (see for instance http://advances.sciencemag.org/content/1/5/e1400253.full). Whether tipping points exist in the biosphere, when they could be reached and what their consequences may be for agriculture (e.g., pollination of crops, plagues), public health (e.g., vector-borne disease), etc., is still unknown.

This is the context of this doctoral dissertation, through which I hope to contribute to closing the gap between two areas of fluvial science that have historically viewed and influenced rivers in fundamentally different ways.

Nonetheless, as I try to show here, methods that were developed for purely engineering purposes can also be used to assess the effects of engineering measures on environmental quality. Furthermore, if applied properly, they can provide ecologically relevant information that is not normally available using the methods commonly employed in ecology, and for which, at the same time, there is no rationale in engineering.

I therefore hope this thesis serves as an incentive for more interdisciplinary work, despite the differences in technical language, goals, views of nature and academic background.

Finally, I would like to thank all the people who supported me during my years as a doctoral candidate. I especially would like to thank and recognize Prof. Stamm's openness and guidance, and of course his patience in supervising a biologist writing a doctoral thesis in engineering.

I would also like to thank PD Dr. Jochen Koop for his guidance during my stay as a guest researcher at the Federal Institute of Hydrology's Animal Ecology Department, and for co-supervising this dissertation.

I must express my deep gratitude to Prof. Daniel Hering and all his colleagues at the University of Duisburg-Essen's Aquatic Ecology department, who kindly shared their research results with me, showing a true commitment to transparency, scientific rigor and open collaboration. I hope to be able to reciprocate in the near future.

I am equally indebted to Prof. Thomas Berendonk and his colleagues at the TU Dresden's Institute of Hydrobiology. Their input concerning the ecological aspects of this dissertation was fundamental.

I would also like to thank Prof. Emil Dister for his sincere interest in this thesis, and for helping me find the right path during the early stages of the PhD, which tend to be the most difficult.

Special thanks also go to all my colleagues at the TU Dresden's Institute of Hydraulic Engineering, from whom I learned a great deal on a scientific and human level. I am also indebted to Prof. Christian Bernhofer and Dr. Sabine Hahn-Bernhofer, and to all the people in the BMBF's IPSWaT scholarship program.

Lastly, I want to thank my family; I owe everything I know and am to them.

This last line is dedicated to my wife. You are in everything I do. Always.

ABSTRACT

A methodological framework is proposed for examining hydromorphological quality using habitat dynamics indicators. The goal of the framework is to provide a judgment of hydromorphology that is as ecologically based and objective as possible. To achieve this, relevant concepts in different areas of ecology are combined with the potential of hydrodynamic numerical models for ecological analysis in order to derive the basic elements of the methodological framework.

Three elements were derived. The first situates the notion of hydromorphology, as derived from the EU Water Framework Directive (WFD), in the conceptual scaffolding of current lotic ecology and fluvial hydraulics. The hydromorphological template is proposed as an alternative, more dynamic concept, which allows viewing hydromorphology not just as a snapshot description of structural features, but rather as the prevailing regime of spatiotemporal variability in substrate and hydrodynamics.

At the same time, hydromorphology's influence is understood as part of a larger network of direct and indirect effects, of which only the direct influence of habitable space, i.e., areas of suitable substrate and hydrodynamics, can be considered in the context of hydromorphological quality analysis. On this basis, limiting- rather than central-type responses are evaluated in the analysis of the hydromorphological template.

The second element deals with the scales of analysis and the variables used to describe the hydromorphological template. The idea of 'ecological neighborhoods' (Addicott et al. 1987) is used to determine the adequate extent and resolution of analysis, which forces spatial and temporal considerations to follow what is known about the basic biology of the target species.

The derivation of the analysis variables follows the 'patch dynamics view' of lotic ecosystems (Pringle et al. 1988, Townsend 1989), which, in line with the physical habitat template concept (Poff and Ward 1990) and habitat template theory (Southwood 1977), emphasizes the dynamic nature of running water environments and the importance of spatiotemporal variability in resources as a central aspect of habitat-biota relationships.

The third and last element of the framework provides a basis for interpreting the values of the patch dynamics indicators derived in the previous step. The guiding principle for such interpretation is the idea that the 'time course of habitat availability and persistence relative to specific life cycle requirements' (Poff and Ward 1990) is fundamental to the survival chances of lotic populations at a site.

The biological basis of this interpretation is tested in a proof-of-concept for the proposed method, which was conducted using available abiotic and macrozoobenthos data. Here, the potential limiting effect imposed by the scarcity of aquatic habitable space on sensitive macrozoobenthos taxa was explored. Resistent taxa were used as control. The limiting effect of aquatic habitable space availability was confirmed, which can be considered as a successful proof of the rationale underlying the proposed approach.

A proposal for the technical implementation of the above analysis scheme is presented at the end of the thesis in the form of R scripts.

KURZFASSUNG

In dieser Dissertation wird ein methodischer Rahmen zur Untersuchung der auf das Makrozoobenthos bezogenen hydromorphologischen Qualität von Fließgewässern anhand von Habitatdynamik-Indikatoren vorgeschlagen. Ziel dieses methodologischen Rahmens ist es, eine möglichst objektive und ökologisch fundierte Bewertung der hydromorphologischen Situation zu liefern, wofür relevante Konzepte aus verschiedenen Bereichen der Ökologie mit geeigneten hydrodynamisch-numerischen Modellierungsmethoden kombiniert werden.

Das so abgeleitete methodologische Konzept besteht insgesamt aus drei Elementen. Das erste Element befasst sich mit der Positionierung des auf der Richtlinie 2000/60/EG des Europäischen Parlaments und des Rates (sog. EG-Wasserrahmenrichtlinie) beruhenden Begriffs „Hydromorphologie" in der konzeptuellen Struktur der modernen Gewässerökologie.

Hierbei wird Hydromorphologie als Teil des „Physical Habitat Template" von Fließgewässern betrachtet. Dies ermöglicht es, Hydromorphologie nicht als die vorliegende strukturelle Konfiguration eines Gewässers zu einem bestimmten Zeitpunkt, sondern als das vorherrschende raumzeitliche Variabilitätsregime zu beschreiben bzw. zu untersuchen. Ein zentraler Punkt dieser alternativen Interpretation der Hydromorphologie ist ihre Betrachtung als limitierender Faktor statt als Prädiktor der biologischen Effekte auf Gemeinschafts -bzw Populationsebene.

Im zweiten Element werden die Variablen und Skalen (räumliche und zeitliche Auflösung sowie Ausdehnung) erarbeitet, die für die Untersuchung des o. g. Variabilitätsregimes mit Bezug auf das Makrozoobenthos nötig sind. Als zentrale Habitatvariable für die dynamische Beschreibung der Hydromorphologie wird der aquatische besiedelbare Raum identifiziert, welcher als Funktion der Verfügbarkeit an geeigneten Verhältnissen von Fließgeschwindigkeit, Wassertiefe und Sohlsubstrat definiert wird.

Die Habitateignung wird in dieser Dissertation anhand von Toleranzintervallen beschrieben, die auf dem bestehenden Basiswissen über die Zielorganismen beruhen. Basierend auf dem „Patch Dynamics View" von Fließgewässern werden anschließend verschiedene Habitatdynamik-Indikatoren vorgeschlagen, die aus den Ergebnissen von zweidimensionalen hydrodynamisch-numerischen Simulationen in einer geeigneten Softwareumgebung (die statistische Sprache „R") berechnet werden können.

Das dritte Element liefert die Auswertebasis für die o. g. Dynamikindikatoren. Das Leitprinzip dieser Auswertung ist durch den Gedanken geprägt, dass der zeitliche Ablauf der Habitatverfügbarkeit, bezogen auf die spezifischen Ansprüche der Zielarten, fundamental für die Überlebenschancen der Zielpopulationen in der untersuchten Flussstrecke ist.

Daran anschließend werden die Ergebnisse eines „Proof-of-Conceptes" zur vorgeschlagenen Methodik zusammengefasst. Anhand vorhandener abiotischen und Makrozoobenthosdaten zur Lahn (in Hessen) konnte die aus der konzeptuellen Basis abgeleitete limitierende Verbindung zwischen den erarbeiteten Habitatdynamik-Indikatoren und der Makrozoobenthos-Besiedlung bestätigt werden.

Taxa mit höherer Resistenz gegenüber Habitatverlusten erreichten höhere mittlere Populationsdichten als sensiblere Arten, welche bei geringer raumzeitlicher Habitatverfügbarkeit limitiert waren. Mit den erzielten Ergebnissen konnte bestätigt werden, dass die vorgeschlagene Habitatmodellierungsstrategie nicht nur eine konzeptuelle, sondern auch eine beweisbare ökologische Basis besitzt.

Ein erster Vorschlag zur technischen Implementierung der entwickelten Methode in der opensource Programmiersprache R ist abschließend in der Dissertation enthalten.

CONTENTS

LIST OF FIGURES

LIST OF TABLES

LIST OF SYMBOLS

Symbol	Description
A_t	Total habitable area in the reach at time t
$\overline{A}$	Time average area of patch i
A_i	Area of patch i
$A_{i,t}$	Area of patch i at time step t
$A_{i,t+\Delta}$	Area of patch i at time step $t + \Delta$
avgDur	Average duration of a patch
avgIntDur	Average duration of inter-spell periods for a patch
avgN	Average local population density [ind./area unit]
ct	Cölbe trained (sampling site)
cr	Cölbe restored (sampling site)
c	Connectivity measure. Probability that a random hypothetical journey starting within a habitable space patch also ends at a suitable point
CONST_15	Colwell's constancy (Colwell 1974) of mean daily discharge of the antecedent 15 years
c_R	Damping factor in Bezzola's (2002) method
cv	Coefficient of variation
CV_4	Coefficient of variation of mean daily flow in the antecedent 4 years
d_i	Duration of patch i
ECDF	Empirical Cumulative Distribution Function
$f_{current}$	Microhabitat suitability relative to current in model Streambugs 1.0 (Schuwirth and Reichert 2013)
$f_{substrate}$	Microhabitat suitability relative to substrate in model Streambugs 1.0 (Schuwirth and Reichert 2013)
f_{temp}	Microhabitat suitability relative to temperature in model Streambugs 1.0 (Schuwirth and Reichert 2013)
GIS	Geographic Information System
h_{dens}	Half-saturation density of biomass under optimal habitat conditions in model Streambugs 1.0 (Schuwirth and Reichert 2013)
h	Water depth
HSNum_15	Number of high flow pulses in the antecedent 15 years
hsAv	Indicator of habitable space availability (see section 8.2.6)
K_{dens}	Self-inhibition term in model Streambugs 1.0 (Schuwirth and Reichert 2013)
logAlDens2	Log-transformed alien species density
logSPDens	Log-transformed ratio of the number of species to the total number of individuals in a sample
lt	Ludwigshütte trained (sampling site)
lr	Ludwigshütte restored (sampling site)

I_Δ Indicator of reach-scale habitable space losses
maxDur Maximum duration of a patch
maxIntDur Maximum duration of inter-spell periods for a patch
MDF_4 mean daily flow of the antecedent 4 years
medianIntDur Median duration of inter-spell periods for a patch
minDur Minimum duration of a patch
minIntDur Minimum duration of inter-spell periods for a patch
MZB Macrozoobenthos
N_t Number of patches at time t
numberSpells Number of existence spells into which the existence of a patch is fragmented
ns_i Number of existence spells for patch i
p-val p-value of regression ANOVA
PREDICT_15 Colwell's predictability (Colwell 1974) of mean daily discharge of the antecedent 15 years
qB discharge at gauge Biedenkopf [m^3/s]
qS Discharge at gauge Sarnau [m^3/s]
resVal Indicator of taxon resistance to mechanical stress
r-sq determination coefficient
SPR Fish species richness
$s_{i0.5}$ Median of existence spell duration for patch i
$s_{i\,sd}$ Standard deviation of existence spell duration for patch i
$s'_{i0.5}$ Median of disappearance period duration for patch i
$s'_{i\,sd}$ Standard deviation of disappearance period duration for patch i
se standard error of estimate, computed as $\sqrt{1/(n-1)\sum(o_i-e_i)^2}$, with o_i = ith observed value, e_i = ith estimated value, n = number of observations
T Total number of time steps in the analysis
$\overline{u}$ Mean velocity along main flow axis
$\overline{u'v'}$ Stream-wise-vertical turbulent shear stress per unit mass
$\overline{u'u'}$, $\overline{v'v'}$, $\overline{w'w'}$ Turbulence intensity per unit mass in the streamwise, horizontal and vertical directions
U_* Shear velocity
v Factor used in calculation of I_Δ
xCbar Time average x coordinate of patch centroid
yCbar Time average y coordinate of patch centroid
WFD EU Water Framework Directive
wt Wallau trained (sampling site)
wr Wallay restored (sampling site)
y_R Thickness of the roughness sublayer
y_W Thickness of the inner region
y Vertical coordinate
ZeroDays_4 mean number of zero-flow days per year in the antecedent 4 years
Δ Time interval for area loss computations
κ von Kármán constant (0.41)
τ Total shear stress
τ_t Turbulent shear stress

1 MOTIVATION

This thesis was stimulated primarily by the following occurrences:

- Hydromorphology is believed to be one of the main remaining obstacles to the achievement of a 'good ecological status/good ecological potential' (GES/GEP) in European river systems (EEA 2012a, Vaughan et al. 2009). This highlights the importance of contributions from hydraulic engineering and river morphology to the achievement of environmental goals in river management, as attested by the growing literature on eco-hydraulics and the importance allocated to hydromorphology in the EU Water Framework Directive (WFD) (EC 2000).

- The fact that, in general, morphological restoration measures have not led to the expected recovery of biological communities (Bernhardt et al. 2005, Haase et al. 2013, Jähnig et al. 2011, Palmer et al. 2010, Sundermann et al. 2011) suggests that the current framework for hydromorphological assessment and restoration must be further developed.

- Given the difficulties involved in constructing empirical models for biota-hydromorphology relations (Downes 2010, Hurlbert 1984, Lancaster and Downes 2010b, Strong 1980, Underwood 1989, 1993, Wiens 2002), quality assessment approaches derived from the accumulated ecological knowledge in this field might provide a useful alternative.

- There exists a division between the bodies of knowledge of stream restoration, stream ecology and stream hydro- and morphodynamics. This division has implied that concepts and methods in each discipline that could potentially be applied in the others have remained unexplored by practitioners. In particular, the capacity of shallow-water hydrodynamic numerical models to provide spatially and temporally explicit habitat information seems to open new possibilities for linking hydromorphology and lotic populations.

The following paragraphs expand on each of the above points.

The dire structural and ecological condition of European river water bodies is summarized in several reports published in the context of management under the WFD (e.g, EEA 2012a,b). Figure 1.1a shows that nearly half the EU's surface water bodies are affected by hydromorphological pressures and are impacted by habitat alteration.

Most of these pressures consist of morphological alterations and flow manipulation ('water flow regulation' in figure 1.1b), which are directly linked to the intensive human use of these surface waters for navigation, hydro-power generation and flood control. In an ecological management context, these modifications can be interpreted as the maximization of a group of these systems' environmental services, coupled with the restriction of the remaining subset of their functions (Tockner et al. 2011).

Germany is one of the EU countries with the highest degrees of hydromorphological intervention (figure 1.2). According to the internet portal of the Umweltbundesamt (consulted on June 25, 2014), only 8% of the 9.070 German running water bodies (rivers and smaller streams) achieved either a 'good' or 'high' ecological status (figure 1.1). The most common causes for this are channelization, dredging and maintenance, impoundment (weirs) and excessive nutrient inputs from agriculture.

Unlike for water quality in the past, removing hydromorphological hindrances to the structural and functional recovery of lotic ecosystems has proven to be a challenging task given the complexity of ecological-physical linkages (Vaughan et al. 2009). Several empirical (Bernhardt et al. 2005, Haase et al. 2013, Jähnig and Lorenz 2008, Jähnig et al. 2010, 2009, Palmer et al. 2010) and theoretical (Bond and Lake 2003, Lake et al. 2007) studies have shown that localized morphological restoration may not be enough to re-establish the desired target communities.

Figure 1.3 shows the change in invertebrate and ground beetle taxa richness in restored relative to trained river reaches in the central German highlands (Jähnig et al. 2009). In this study, success, measured as an increase in taxa richness in restored reaches, was restricted to certain groups

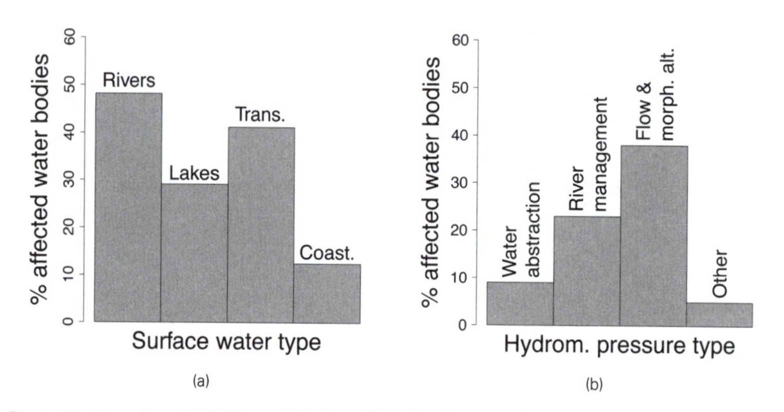

Figure 1.1: Proportion of (a) EU water bodies affected by hydromorphology-related pressures in general ("Coast." = coastal waters, "Trans." = transitional waters) and (b) EU river water bodies affected by specific types of hydromorphology-related pressures ("Flow & morph. alt." = flow regulation and morphological alterations. Re-drawn using data from EEA (2012b).

Table 1.1: Percentage of German running water bodies in the five ecological status categories. Re-drawn using data from the German Environment Agency (Umweltbundesamt, http://www.umweltbundesamt.de/, consulted on June 25, 2015.

Ecological status of running waters in Germany*		
Ecological Status Class	**Ecological Status**	**Percentage**
1	High	0.1
2	Good	7.7
3	Moderate	29.1
4	Poor	35.8
5	Bad	24.4

*% of water bodies in the individual classes. 2.8% not evaluated. Total number of water bodies = 9070 as of 22.03.2010.

(ground beetles, aquatic macrophytes). According to recent reviews (e.g., Jähnig et al. 2011), this seems to be related to factors such as the dispersal capacity of colonizing taxa, time after restoration and ecological quality of the neighboring stretches of the stream network (see below). Similar results have been reported for this region by Jähnig and Lorenz (2008) (figure 1.4).

The reasons for restricted restoration success are manifold. Recent accounts (e.g., Didderen and Verdonschot 2010, Hughes 2007) have found evidence that points to the importance of processes such as dispersion, recolonization and landscape-scale processes. Jähnig et al. (2011), for instance, indicate the problem that restoration measures be commonly carried out on stretches

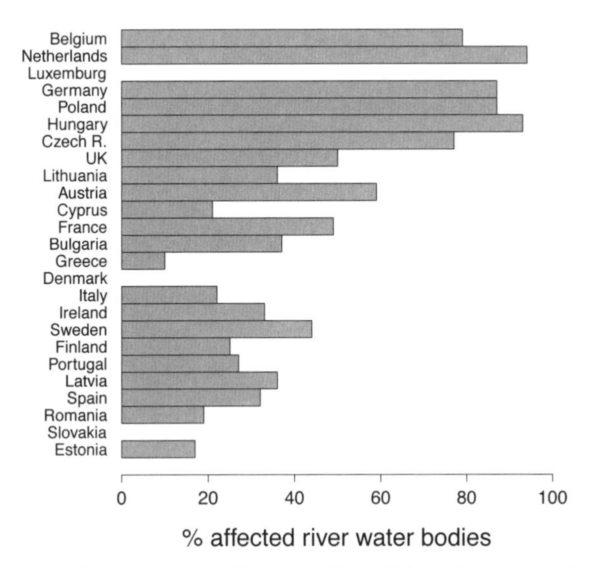

Figure 1.2: Percentage of river water bodies affected by hydromorphology-related pressures in different EU member states. Member states are arranged from top to bottom in increasing proportion of water bodies in 'good' or better ecological status/potential. Empty bars correspond to missing data. Re-drawn using data from EEA (2012a).

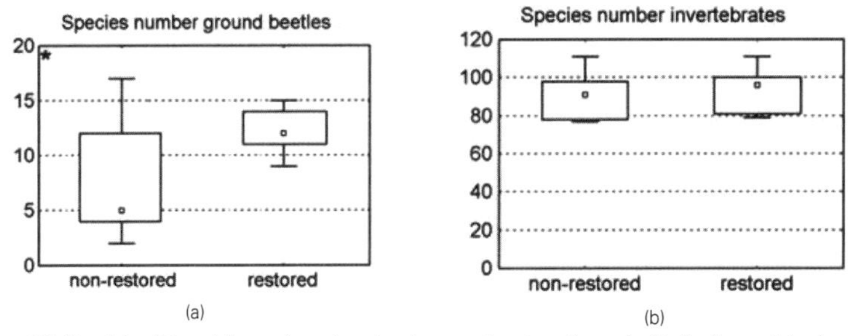

Figure 1.3: Boxplots of taxa richness in restored and non-restored sections of mid-sized mountain rivers in Germany reported by Jähnig et al. (2009). (a) Ground beetles, (b) aquatic invertebrates. White square = median, box = 25–75% range; whiskers = min–max; * = differences are significant at $P < 0.05$ (U-test).

no longer than a few hundred meters and are wrongly expected to enhance river water bodies up to several kilometers in length.

In their review, Bond and Lake (2003) also address this issue, highlighting the importance of 'long-term and large-scale processes' and 'inappropriate scales of restoration' among the factors responsible for the low effectiveness of many restoration attempts (see table 1.2).

These shortcomings in the practice of stream restoration are related to incomplete knowledge of the relations between physical habitat and stream populations. Empirical approaches to this problem typically lead to statistical habitat-biota relations with a degree of scatter that hinders their use as predictive tools. Moreover, there can be many ecological reasons for this so-called 'statistical noise', which may be summarized as follows.

Firstly, as seen in figure 1.7, unmeasured factors may have just as important an effect as those under study. In fact, understanding scatter around regression curves in exploratory studies as 'noise' is illustrative of the common underestimation of multiple causality that is inherent to natural systems (Downes 2010).

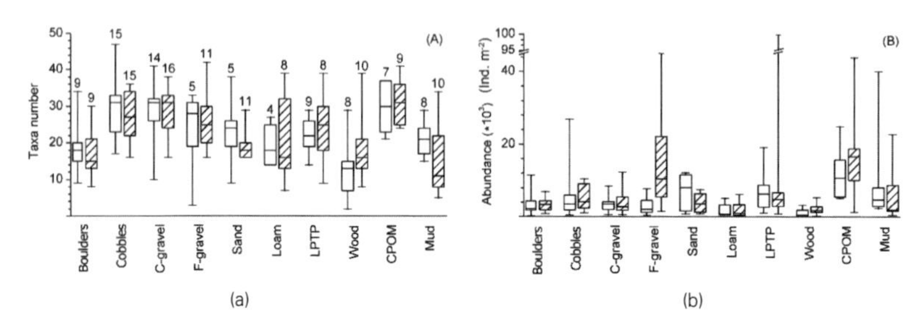

Figure 1.4: Box-and-Whisker plots of (a) taxa number and (b) abundance reported by Jähnig and Lorenz (2008) for specific substrate types in restored (multiple-) and non-restored (single-channel) sections of mid-sized mountain rivers in Germany.

Table 1.2: Important ecological questions to consider in the planning and target setting stages of habitat restoration. Source: Bond and Lake (2003).

---

1. Are there barriers to colonization?
    - What and where are the source populations?
    - How can potential barriers be overcome?
2. Do the target species have particular habitat requirements at different life stages?
    - What are these requirements?
    - How should these habitats be arranged spatially?
3. Are there introduced species that may benefit disproportionately to native species from habitat restoration?
    - Can colonization of these organisms be restricted?
4. How are long-term and large-scale phenomena likely to influence the likelihood, or timeframe of responses?
    - Will these affect the endpoints or just the timeframe of responses?
    - How will this affect monitoring strategies, and can monitoring strategies be adjusted to deal with this?
5. What size habitat patches must be created for populations, communities and ecosystem functions to be restored?
    - Is there a minimum area required?
    - Will the spatial arrangement of habitat affect this (e.g. through the outcomes of competition and predation)?

In similar vein, changes triggered by modifications of habitat structure have multiple causal pathways and are by no means exclusively hydraulic/morphological. Hydraulic engineering measures can also affect food availability, water quality, spatial processes and biotic interactions (Schuwirth 2012).

Further, the effect of the studied factors may or may not result in changes in local abundance depending on whether resistance/resilience thresholds are exceeded (Lancaster and Downes 2010a,b). In this sense, limiting response instead of central response analyses should be employed (Lancaster and Belyea 2006) (section 5.1).

Similarly, the distribution and abundance of lotic biota is not the result of optimal selection of suitable habitat according to fixed abiotic preferences produced by adaptation. As is argued in recent discussions on eco-hydraulics, this is a common misconception in the river management literature that has been challenged (e.g., (Anderson et al. 2006, Lancaster and Downes 2010a, Underwood et al. 2004) on the basis of evolutionary theory (Sober 1985, Williams 1974) and experimental studies (Downes et al. 2000).

These considerations make it clear that any attempt to link changes in lotic communities with river training measures immediately faces the problem of multiple causality, i.e., it is not one but a multiplicity of mechanisms that are responsible for the observed biological changes (figure 1.5), only a part of which is directly attributable to engineering measures. Hence, disentangling the influence of the factor(s) of interest is a major challenge in such undertakings.

Further evidence of this phenomenon was found by Kail and Hering (2009), who reported a positive effect of high upstream (up to 2.5 km) morphological quality on local ecological status for

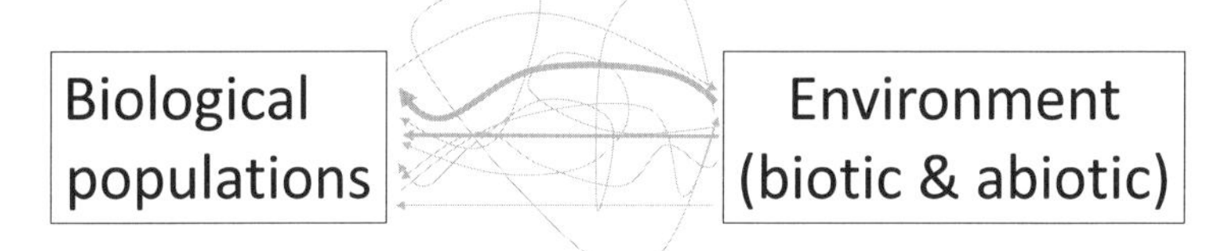

Figure 1.5: Relationships between the WFD relevant biotic populations and their environment can be understood conceptually as a network of complex interactions (arrows). Each of these represents an individual causal pathway or process, and processes may be non-linear and interdependent. Thicker and less curved arrows represent stronger and more direct linkages.

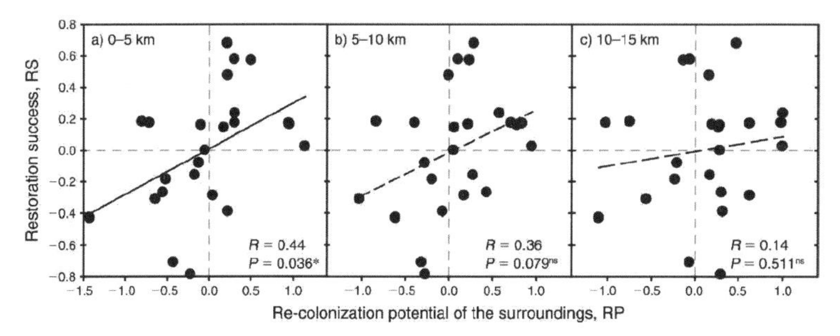

Figure 1.6: Relationship between restoration success (RS) and recolonization potential (RP) of neighboring sites within a radius of (a) 0–5, (b) 5–10, and (c) 10–15 km of the restored location. Source: Sundermann et al. (2011).

sites on streams in central and western Germany. This effect was larger for sites with an intermediate ecological quality, which led these authors to argue that the 'spreading effect' should be understood as the input of individuals from good-quality upstream sites maintaining the ecological status of morphologically poor locations, which seemed incapable of sustaining viable local populations.

A report published by the German Umweltbundesamt (Dahm et al. 2014) summarizes the above points and describes strategies to optimize stream restoration efforts. According to these authors, among the main factors to be taken into account in restoration planning are the existence of source populations in the catchment, the dispersal capacity of the target organisms and related factors, and persisting pollution-related stresses.

Despite the success of current bio-assessment schemes in classifying the ecological status of European running waters, discussions in the freshwater ecology literature (e.g., Bond and Lake 2003, Downes 2010, Lake et al. 2007, Lamouroux et al. 2010) and the ecological literature in general (e.g., Orians 1980) point to the need of expanding the current framework. Issues of scale (Hering et al. 2010) and causal relations between hydraulics, morphology and biological responses (Haase et al. 2013, Lancaster and Downes 2010a,b, Statzner and Bêche 2010) are deemed not yet sufficiently addressed, and are drawn on to explain the aforementioned shortcomings faced by management under the directive.

Further motivation for focusing on process and causality is provided by CIS guidance document No. 3 (IMPRESS 2003). This document lays out a framework for impact assessment, the so-called Driver-Pressure-State-Impact-Response chains, and thus emphasizes the importance of causal relationships for ecological assessment and restoration under the WFD.

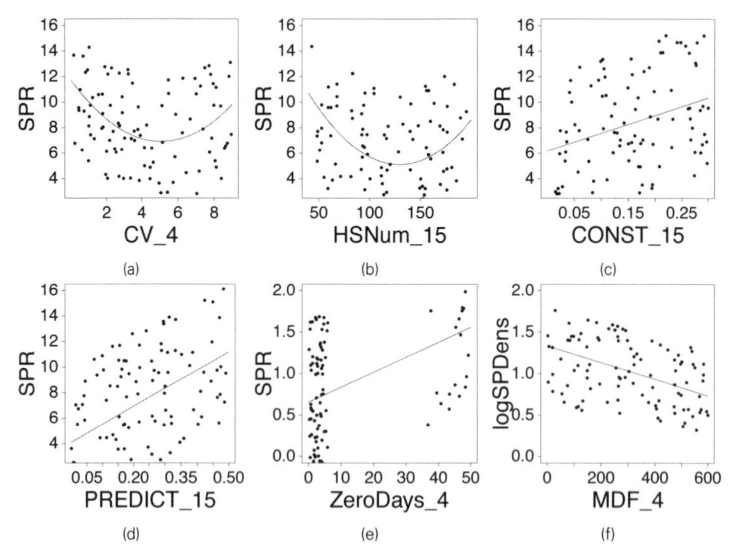

Figure 1.7: Random example of correlation analyses linking physical habitat characteristics and biological responses (fish metrics) in multiple river catchments (Arthington et al. 2012). Plots were re-drawn by adding random noise to the correlation equations reported by Arthington et al. (2012), with the purpose of illustrating the uncertainty (large scatter) of predictions based on this type of analysis. For acronyms in plot axes, see list of symbols.

In response to these difficulties, more process oriented approaches have been proposed that emphasize aspects such as the re-colonization potential of the restored site and the natural dynamics of the stream (Beechie et al. 2010, Dahm et al. 2014). Clear evidence justifying this shift was presented by Sundermann et al. (2011), who found that the surrounding pool of colonists, expressed as the Fauna Index of sites on neighboring streams, is significantly related to restoration success (figure 1.6).

From the point of view of hydraulic engineering, these ideas can contribute to the development of methods that guide training and restoration measures with respect to their biological consequences. At the time of writing of this thesis, different tools, which can collectively be termed 'eco-hydraulic' in nature, are available for simulating habitat quality on the basis of local hydraulics, morphology and habitat preference.

However, a series of advances in river ecology have taken place in the last few decades that are still outside the focus of existing approaches. In fact, spatial population dynamics and riverine landscape ecology have only recently begun to permeate management (e.g., Dahm et al. 2014, Kail et al. 2012, Sondermann et al. 2012), and incorporating advances in these areas into ecohydraulics by building a basic analytic framework could contribute to the advancement of both fields.

All the above issues highlight the need for the development of more process-oriented approaches, which might help bridge the gap between hydromorphology and biology and provide realistic estimations of what can be expected as the result of hydromorphological interventions. The opportunity therefore exists to explore conceptual linkages between these two areas, as well as the potential of hydraulic simulation models, spatial analysis tools, and alternative views of ecosystem functioning in general (e.g., Pickett et al. 2007).

2 OBJECTIVES OF THE DISSERTATION

This dissertation aims to provide a conceptual and technical framework for linking hydromorphology and ecological quality, which is expected to serve as an objective basis for judging the ecological suitability of the physical features of a stream reach. For this purpose, macrozoobenthos is used as biological target group given its relation to hydromorphology and its relevance in the WFD.

The fundamental aim of the thesis is therefore to explore the possibilities offered by alternative approaches to direct empirical correlations based on monitoring data. The resulting approach is presented in the form of a methodological framework consisting of three parts, as described in section 3.

The objectives of this thesis can be summarized as follows:

General objective

- Provide a conceptual and methodological framework that contributes to linking hydromorphology and ecological quality based on the use of shallow water simulation tools at the appropriate scales.

Specific objectives

- Situate hydromorphology in the conceptual structure of modern river ecology by collating relevant concepts and methods in the fields of river ecology, geomorphology and hydrodynamics.

- Develop the basic elements of the methodological framework based on this new perspective of hydromorphology

- Present a first technical implementation of the framework that includes scripts for computing the proposed habitat quality indicators

- Provide a proof-of-concept of the resulting methodological framework using available hydromorphological and macrozoobenthos community data.

3 APPROACH AND STRUCTURE OF THE THESIS

The WFD biological quality component chosen in this thesis was macrozoobenthos (MZB). These organisms are among the most widely used biological groups in ecological assessment under the WFD. This is mainly due to their ubiquity, well-studied taxonomy (excluding certain groups), known habitat relations (Schmidt-Kloiber and Hering 2012) and technical convenience (Fore et al. 1996, Meier et al. 2006).

Further, many taxa in this group exhibit a close association with the physical characteristics of their habitat, especially at the micro-scale (Hering et al. 2006, Poff and Ward 1990, Schröder et al. 2013). This last aspect is particularly relevant for the purposes of this thesis, since it implies that hydromorphology can be expected to play an important role in the lives of many macrozoobenthos taxa.

With this biological target group, a methodological framework was developed for assessing stream hydromorpholgy from an ecological point of view. This was undertaken following the three steps shown in figure 3.1. First, a conceptual basis was constructed (section 5) based on a review of the literature on hydromorhology (chapter 4) and phyical habitat in stream ecology (chapter 5).

In the second step (section 6), variables and scales for the ecological analysis of stream hydromorphology were derived from the conceptual basis. An alternative view of hydromorphology is presented in this context, concluding with a series of indicators for describing this new interpretation of hydromorphology.

This is followed by the technical implementation of the derived indicators, and a discussion on the criteria for conducting the necessary numerical shallow water simulations and the acquisition of the required substrate data (chapter 7).

The final step is described in chapter 8, which presents a proof-of-concept for the proposed approach using available hydromorphological and macrozoobenthos data. This step was undertaken in collaboration with the University of Duisburg-Essen's Aquatic Ecology Department and the Federal Institute of Hydrology's Animal Ecology Department (BfG, Referat U4 – Tierökologie).

This step aims to ensure the ecological validity of the approach by exploring the relationship between the proposed habitat dynamics indicators and the overall population-level success of a group of type-specific target taxa.

Finally, chapter 11 presents an informatic implementation of the approach in the form of scripts written in the programming language R.

3.1 CONSTRUCTION OF A CONCEPTUAL BASIS

This component is based on a review of existing concepts and methods that address hydromorphology and its relationship with aquatic macrozoobenthos. The review comprises methodologies for hydromorphological assessment derived in the context of the WFD (CEN 2004, LAWA 2000, Raven et al. 1998), basic concepts in lotic (e.g, Poole 2002), landscape (e.g., O'Neill et al. 1988, Turner et al. 2001) and spatial (Hanski 1999b, Leibold et al. 2004, Ranta et al. 2006) ecology, and eco-hydraulic modeling approaches of varying complexity (e.g., Kopecki and Schneider 2010, Schuwirth and Reichert 2013).

In order to clarify the role of hydromorphology as an abiotic factor in the ecology of stream macrozoobenthos, the factor-group framework proposed by Schuwirth (2012) is used (section 5.1).

This step aims to extract key ideas from the available body of knowledge in order to construct a conceptual basis that helps addressing the gap between hydromorphology and biology. The result of this synthesis is a set of principles that are used as a guide in the derivation of the basic elements of the approach.

These principles draw on concepts developed in the aforementioned areas of ecology in the last few decades, many of which are recognized in recent discussions on eco-hydraulics as a potential strategy for the advancement of this field (Lancaster and Downes 2010b, Vaughan et al. 2009).

In the context of this review, key concepts for the derivation of the methodological framework are identified. Habitat template theory (Southwood 1977), the importance of temporal and spatial heterogeneity in streams (Pringle et al. 1988, Townsend 1989) and the concept of the physical habitat template for running waters (Poff and Ward 1990) are used as a basis for developing the method.

3.2 DEVELOPMENT OF THE METHODOLOGICAL FRAMEWORK

This component builds on the conceptual basis established in the previous step. It aims to develop a description of hydromorphölgy that is more closely related to the way the long-term

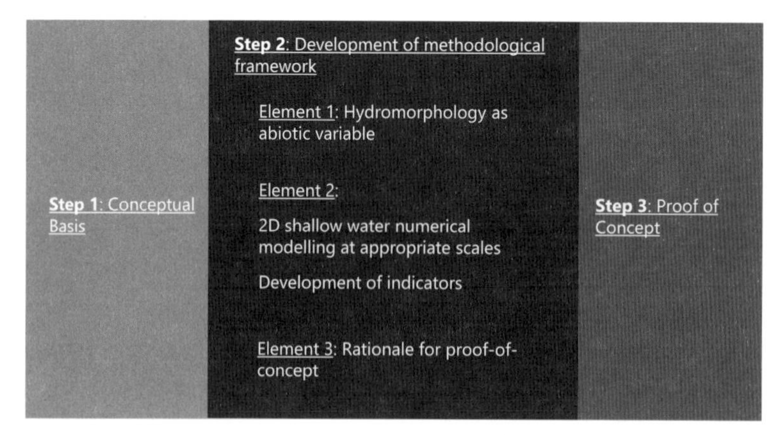

Figure 3.1: Schematic representation of the approach followed in this dissertation.

survivorship of the target organisms is related to their environment. The development of the method is structured into three elements.

In the first element, a re-interpretation of hydromorphology in the context of the conceptual basis is presented. This re-interpretation allows deconstructing hydromorphology into primary variables that bear a more close relationship to the long-term survival of macrozoobenthos in streams.

At the same time, hydromorphology is placed in the larger ecological context of all the factors that influence macrozoobenthos persistence, which allows identifying its role as a limiting factor rather than as a predictor.

This provides the basis for the second element, in which a strategy for implementing numerical two-dimensional (2D) shallow water models on an ecologically sound basis is devised. This strategy allows setting up the models at temporal and spatial scales that are relevant to the ecological process at hand, namely the long-term survivorship of the target species.

The results of these simulations can then be summarized into a series of indicators that are related to the long-term survivorship of stream macrozoobenthos according to the conceptual basis.

The third and last element describes how the ecological basis of the approach can be confirmed using available macrozoobenthos and hydromorphological data, which is presented in chapter 8. This confirmation is not intended as a proof of the universal validity of the approach, but rather as a proof-of-concept that provides certainty to its underlying rationale.

3.3 TECHNICAL ASPECTS

This section describes the technical implementation of the proposed approach. The description of hydromorpholgy proposed here is based on results from 2D shallow water numerical simulations, substrate mapping and knowledge of the biology of the target species in the form of habitat association rules.

These three components are linked by means of scripts written in the statistical language R. The scripts use R's spatial analysis capabilities in order to calculate the habitat indicators described in section 6.2.

3.4 PROOF-OF-CONCEPT

The purpose of this section is, first, to explore the possibilities and limitations of the proposed methodological framework, and second, to carry out an empirical test of its underlying rationale.

As described in chapter 8, this proof-of-concept aimed to make use of the detailed habitat information provided by the proposed approach, which was used in order to conduct a hypothesis test on the limiting role of hydromorphology following the control species approach (Downes 2010).

This allowed proving the hypothesis that a limiting effect would be detected for taxa expected to be more sensitive to habitable space scarcity, whereas such limitation would be much weaker for more tolerant taxa. These results provide a first empirical check of the rationale behind the method, and at the same time a confirmation of the ecological and fluvial hydraulics principles on which it is based.

The proof of concept was carried out using biological and hydromorphological data available for the river Lahn (Federal State of Hessen, Germany). The biological and micro-habitat data were kindly provided by the University of Duisburg-Essen's Aquatic Ecology Department, stemming from studies published by Januschke et al. (2014) and (Jähnig et al. 2008) on the temporal differences of macrozoobenthos communities of restored and trained sites.

The abiotic information (reach geometry, aerial photography, discharge time series, etc.) was kindly supplied by the Hessische Verwaltung für Bodenmanagement und Geoinformation and the Hessisches Landesamt für Umwelt und Geologie. The trait database for macrozoobenthos used in the computation of taxa resistence values was provided by the German Federal Institute of Hydrology (Bundesanstalt für Gewässerkunde, BfG). These data are thoroughly described in section 8.2.1.

4 HYDROMORPHOLOGY IN THE CURRENT STATE OF THE ART

4.1 AREAS FOR POTENTIAL DEVELOPMENT

One of the most important tenets of the WFD is the idea that the ecological quality of river systems should be assessed on the basis of their biological communities, specifically their taxonomic and functional composition. In this context, hydromorphological restoration measures can only be considered successful when they lead to the recovery of type-specific biota, having as reference either natural (Ecological Status) or modified (Ecological Potential) conditions (EC 2000).

Thus, impact assessment and restoration planning require methods that allow predicting the level of biological recovery that can be expected under the existing boundary conditions. The need for such planning tools is further emphasized by several field (Haase et al. 2013, Jähnig et al. 2010, Palmer et al. 2010, Sundermann et al. 2011) and theoretical studies (e.g., Bond and Lake 2003, Lake et al. 2007), which have shown that local morphological restoration alone does not necessarily lead to biological recovery.

Therefore, attempts to incorporate further ecological considerations into the analysis of hydromorphology are necessary in the current management context. Developments in community and spatial ecology in the past decades may offer useful reference points for this task, particularly in regards to improving current perceptions of hydromorphological quality.

One of the central points in this context is that hydromorphology should be described in dynamic rather than static terms (Palmer et al. 1997). As will be explained in the coming chapters, the approach presented in this thesis does not concentrate exclusively on whether habitat at a given point in space and time is good or bad for a particular species.

Rather, it uses knowledge on its habitat associations to estimate the spatiotemporal distribution of its **'aquatic habitable space'** (Downes and Reich 2008) at appropriate spatial and temporal scales (see section 6.1), and then uses this distribution to draw conclusions about the hydromorphological quality of the reach.

Extending focus beyond what is 'good' or 'bad' habitat locally and at a particular moment helps avoiding the shortcomings of static descriptions of hydromorphology (Beechie et al. 2010, Palmer et al. 1997). In the case of restoration, such methods can lead to project outcomes that may benefit habitat structure for the target organisms but whose overall effect could be detrimental because of their negative influence on related food web components (Leal 2012).

4.2 ADDRESSING THE GAP BETWEEN BIOLOGY AND HYDROMORPHOLOGY

4.2.1 Hydromorphology in the context of the WFD

The WFD does not provide an explicit definition of hydromorphology. However, its meaning in this context can be gathered from the variables listed as hydromorphological quality elements in Annex V (EC 2000):

- Hydrological regime
  - Quantity and dynamics of water flow
  - Connection to groundwater bodies
- River continuity
- Morphological conditions
  - River depth and width variation
  - Structure and substrate of the river bed
  - Structure of the riparian zone

Other sources that contribute to the current framework do provide a definition. According to the European Committee for Standardization (CEN 2004), hydromorphology refers to the 'physical and hydrological characteristics of rivers including the underlying processes from which they result'.

Although some discussion may still exist regarding its exact meaning (e.g.,Vogel 2011), there is general agreement on hydromorphology's central aspects. The European Environment Agency (EEA 2012b), for instance, recognizes as hydromorphological pressures:

- hydrological alterations, such as changes in the flow regime due to impoundment (weirs, barrages, sluices, locks, dams) or water abstractions/diversions,

Table 4.1: WFD definitions for high, good and moderate ecological status (EC 2000).

Element	High status	Good status	Moderate status
General	There are no, or only very minor, anthropogenic alterations to the values of the physico-chemical and hydromorphological quality elements for the surface water body type from those normally associated with that type under undisturbed conditions. The values of the biological quality elements for the surface water body reflect those normally associated with that type under undisturbed conditions, and show no, or only very minor, evidence of distortion. These are the type-specific conditions and communities.	The values of the biological quality elements for the surface water body type show low levels of distortion resulting from human activity, but deviate only slightly from those normally associated with the surface water body type under undisturbed conditions.	The values of the biological quality elements for the surface water body type deviate moderately from those normally associated with the surface water body type under undisturbed conditions. The values show moderate signs of distortion resulting from human activity and are significantly more disturbed than under conditions of good status.

- morphological alterations of the banks, bed or channel planform due to engineering measures (revetment, levees, channelization, cutoff, dredging, groynes, longitudinal dykes) and/or land use change in the catchment, and
- floodplain disconnection and land reclamation.

Existing methodologies for hydromorphological quality assessment also provide details of what is understood as hydromorphology in the current framework. For instance, a comprehensive definition is provided by LAWA (2000). In this method, the term 'water structure' (in German: 'Gewässerstruktur') is employed, being defined as '(. . .) all spatial and material differentiations of the stream bed and its surroundings as far as they are hydraulically, stream-morphologically and hydrobiologically effective and relevant to the ecological functions of the stretch of water and the floodplain'.

In this dissertation, an operational definition of hydromorphology is adopted, which was derived by introducing explicit scale considerations to the definition proposed by CEN (2004). The definition adopted in this thesis is as follows:

> *hydromorphology refers to the morphological and hydrodynamic aspects of stream habitats, observed at the spatial and temporal scales at which the processes that influence population persistence operate.*

4.2.2 Hydromorphological quality in the context of the WFD

Hydromorphological quality assessment in the current framework is based primarily on a comparison with low-impact (ideally impact-free) reference conditions. This is evident in the methods available for determining hydromorphological quality (CEN 2004, LAWA 2000, Raven et al. 1998), as well as in the WFD definitions for high, good and moderate ecological status (table 4.1) and potential (table 4.4).

The scoring systems used to determine hydromorphological quality in these methods employ type-specific reference conditions as quality benchmarks, for also which type-specific hydromorphologically unimpacted sites are used as a guideline.

Table 4.2: Summary of variables for hydromorphological assessment employed in CEN (2004).

Channel	*Geometry* – planform, longitudinal section, cross-section; *Substrates* – artificial, natural, management/catchment impacts; *Vegetation and organic debris* – structural form of macrophytes present, leafy and woody debris, vegetation management; *Erosion/deposition character* – features in channel and at base of bank; *Flow* – flow patterns, flow features, discharge regime; *Longitudinal continuity as affected by artificial structures* – artificial barriers affecting continuity of flow, sediment transport and migration for biota
Banks/riparian zone	*Bank structure and modifications* – bank materials, types of revetment/bank protection; *Vegetation type/structure on banks and adjacent land* – structure of vegetation, vegetation management, types of land use, extent and types of development
Floodplain	*Adjacent land use and associated features* – types of land use, extent and types of development, types of open water/wetland features; *Degree of (a) lateral connectivity of river and floodplain; (b) Lateral movement of river channel* – degree of constraint to potential mobility of river channel and water flow across floodplain, continuity of floodplain

Thus, at a given sampling site, the hydromorphological variables making up the assessment (summary in tables 4.2, 4.5 and 4.6) are assigned value based on the degree of modification from these reference conditions.

In the methodology by LAWA (2000), for instance, the reference is provided by the so-called 'present-day potential natural state', which is the 'state that would establish itself after abandoning of all existing uses in and on the water and its floodplain and after the removal of all buildings there'.

Following this principle, assessment variables are assigned an integer value from 1 to 7 depending on the site's deviation from the reference. European Standard 14614 (CEN 2004) proposes as scoring principle the degree to which assessment features meet criteria for type-specific reference conditions with respect to 1) bed and bank character, 2) planform and river profile, 3) lateral connectivity and freedom of lateral movement, 4) free flow of biota, water and sediment in the channel, and 5) vegetation in the riparian zone.

Lastly, scoring in the River Habitat Survey method (Raven et al. 1998) is based on the presence of channel and river corridor features which are 'known to be of value to wildlife', with 'diversity and naturalness of physical structure' being the two main criteria determining habitat quality. In all cases, professional judgment is a central element of the assessment process.

Despite the widespread use of these methods, the way reference conditions are set is still a matter of debate. Recent discussions (e.g, Rinaldi et al. 2013, Wyżga et al. 2012) advocate for

Table 4.3: Examples of indicators and their scores in the Morphological Quality Index (Rinaldi et al. 2013). GF = Geomorphological functionality, A = Artificiality, CA = Channel Adjustment.

Type	Indicator	Class	Score
GF	F1	A - Absence of alteration in the continuity of sediment and wood	0
		B - Slight alteration (obstacles to the flux but with no interception)	3
		C - Significant alteration (complete interception of sediment and wood)	5
	F2	A - Presence of a continuous (>66% of the reach) and wide floodplain (>n·W, where n=1 or 2 for wandering - braided or for single thread channels, respectively, and W=channel width)	0
		B - Presence of a discontinuous (10-66%) floodplain of any width or >66% but narrow	3
		C - Absence of a floodplain or negligible presence (≤10% of any width)	5
A	A1	A - No significant alteration (≤10%) of channel-forming discharges and Q with return interval >10 years	0
		B - Significant alteration (>10%) of Q with return interval >10 years	3
		C - Significant alteration (>10%) of channel-forming discharges	6
CA	CA1	A - Absence of changes in channel pattern from 1950s	0
		B - Change to a similar channel pattern from 1950s (partly confined - unconfined) or change of channel pattern from 1950s (confined)	3
		C - Change to a different channel pattern from 1950s (only partly confined - unconfined)	6

definitions of hydromorphological reference conditions that go beyond the degree of human intervention and take into account the dynamic nature of river systems, i.e., their adjustment process to changing boundary conditions. This perspective is summarized in the concept of the 'evolutionary trajectory' of rivers (Dufour and Piégay 2009).

Similarly, 'naturalness', defined as closeness to pristine conditions, has been criticized with the argument that pre-human conditions are unachievable and impractical in many cases. This view therefore proposes that, rather than striving for a return to past configurations, guiding images for restoration should aim to 'move the river towards the least degraded and most ecologically dynamic state possible, given the regional context' (Palmer et al. 2005).

The concept of 'guiding image' (in German: 'Leitbild') proposed in LAWA (2000) already incorporates these views by using 'overall concepts of regional landscapes' (in German: 'naturraumspezifische Leitbilder') as guidance in the scoring process.

This alternative view of reference conditions and restoration goals is reflected in the scoring systems of new methodologies for hydromorphological quality assessment (e.g., Morphological Quality Index; Rinaldi et al. 2013). Even though quality decreases with degree of human modification as in previous methods (table 4.3), the selection of functionality indicators (e.g., channel confinement, field evidence of adjustment, alteration in water, wood and sediment fluxes, among others) allows the method to address key aspects of morphological dynamics, and thus provide a more comprehensive picture of the overall hydromorphological status.

4.2.3 The need for biological criteria in hydromorphological analysis

Biological quality components are a key element of ecological status/potential classification in both the WFD and related hydromorphological asessment methodologies. In LAWA (2000), for instance, it is explicitly stated that structural quality (in German: 'Gewässerstrukturgüte') refers to the ecological functionality (in German: 'ökologische Funktionsfähigkeit') of the river system's physical structure. A similar reasoning is applied by Raven et al. (1998) in defining 'value to wildlife', 'diversity' and 'naturalness' as the main determinants of habitat quality.

Recent methodologies for hydromorphological assessment (e.g., Rinaldi et al. 2013) take account of the dynamic nature of rivers, and thus aim to overcome the difficulties associated with the use of static quality targets (Dufour and Piégay 2009). However, given that biological quality represents a critical component of ecological status and potential under the WFD, quality criteria for hydromorphology should also have explicit and clear causal links to biological responses.

Table 4.4: Definitions of maximum, good and moderate Ecological Potential for the hydromorphological quality elements of heavily modified and artificial water bodies as they appear in Annex V of the WFD (EC 2000).

Maximum ecological potential	Good ecological potential	Moderate ecological potential
The hydromorphological conditions are consistent with the only impacts on the surface water body being those resulting from the artificial or heavily modified characteristics of the water body once all mitigation measures have been taken to ensure the best approximation to ecological continuum, in particular with respect to migration of fauna and appropriate spawning and breeding grounds	There are slight changes in the values of the relevant quality elements as compared to the values found at maximum ecological potential	There are moderate changes in the values of the relevant biological quality elements as compared to the values found at maximum ecological potential. These values are significantly more distorted than those found under good quality

In spite of this, hydromorphological variables and quality standards are still somewhat disconnected from the way organisms perceive their environment, and this disconnect can be identified as one of the more important causes of the continuing low ecological quality of surface waters in Europe (EEA 2012a, Jähnig et al. 2011).

Therefore, from an ecological perspective, hydromorphological quality should reflect the role of the riverine environment's physical structure and dynamics in sustaining viable populations of specific biota groups. This need for a more explicit linkage between biology and hydromorphology has been recognized in the eco-hydraulic literature for some time, particularly by Clarke et al. (2003), Fisher et al. (2007), Lancaster and Downes (2010b), Newson and Large (2006), Vaughan et al. (2009), Wohl et al. (2005).

The above authors highlight the importance of causality in quality assessment and restoration. A central point is that, while it is true that 'naturalness' and degree of adjustment to current boundary conditions can provide good measures of ecological degradation, they do not allow clarifying causal pathways and therefore cannot lead to predictions of the expected effects of physical interventions, either positive (restoration) or negative (degradation).

The difficulties associated with linking hydromorphology and biological responses are manifold and have been described in different ways in the ecological literature. In many cases, observational field studies aiming to quantify the relative importance of hydromorphological factors face methodological difficulties related to 1) mismatches between the scales of observation and the scales at which processes take place (Downes 2010, Wiens 2002), 2) spatiotemporal variability in natural populations (Underwood 1989) and 3) availability of appropriate controls and replication (Hurlbert 1984, Strong 1980, Underwood 1993), among others.

At the same time, the nature of hydromorphology itself, as compared to water quality, has been used to explain the difficulties associated with physical habitat restoration. Whereas the effects of pollution can and have been controlled in many regions by managing point and diffuse sources, hydromorphological restoration requires a wider variety of approaches whose applicability strongly depends on the availability of land (Jähnig et al. 2011).

The idea of 'naturalness' (as proposed in Raven et al. 1998) as a hydromorphological quality criterion can be traced back to the concept of 'biological integrity' (Angermeier and Karr 1994), which 'refers to conditions under little or no influence from human actions'. Under such conditions, biological communities reflect 'natural evolutionary and biogeographic processes'. Nonetheless,

Table 4.5: Summary of variables for hydromorphological assessment employed in Raven et al. (1998).

Predominant valley form: Shallow 'v', deep 'v', gorge, concave bowl, symmetrical floodplain, asymmetrical floodplain, terraces
Number of riffles, pools and point bars
Physical attributes: *Banks* – material, modifications, features; *channel* – substrate, flow type, modifications, features
Banktop land use and vegetation structure: Land use within 50 m of banktop
Channel vegetation types
Bank profiles: *Unmodified* – vertical/undercut, vertical + toe, steep (>45°), gentle, composite; *Artificial* – resectioned, reinforced (whole bank, toe only), artificial two-stage, poached, embanked, set-back embankment
Extent of trees and associated features
Extent of channel features
Channel dimensions
Artificial features
Evidence of recent management
Features of special interest
Choked channel Notable nuisance plant species
Overall characteristics: *Major impacts* – landfill, litter, sewage, abstraction, dam, road, rail, industry, housing, mining, quarrying, overdeepening

there are two limitations to the use of pristine nature as a quality standard in riverine hydromorphology.

First, using 'naturalness' as a quality standard assumes that returning a site along the river corridor to pre-human conditions automatically guarantees the viability of the populations of all target biota groups. This perception bypasses causality, disregards the intricate nature of the mechanisms that determine the persistence of natural populations in a landscape (Boyce 1992, Dunning et al. 1992, Kareiva 1990), and overlooks the fact that nature is inherently variable and does not constitute an idyllic setting where all populations of all species remain viable at all places and times (Pickett et al. 2007, Talbot 1997).

The second limitation is pragmatic. It has been argued that a pristine state does not necessarily constitute a sensible management goal in landscapes that have been extensively modified for centuries, and are likely to remain so in the short- and medium-term (Rinaldi et al. 2013, Wyżga et al. 2012).

In attendance to this, new perceptions in river ecology, such as the concepts of 'domesticated ecosystems' and 'novel communities' (Tockner et al. 2011), have been put forth that recognize 1) the pervasiveness of human influence, especially in industrialized regions, and 2) the necessity to address causality in habitat-biota relations in order to set more realistic quality and restoration standards.

The second limitation is addressed in the current framework, at least in part, through the concept of heavily modified water bodies (EC 2000), by which intensively used rivers are allocated environmental goals (Ecological Potential instead of Ecological Status) in accordance with their modified hydromorphological character. Management and quality determination with respect to invasive species, however, is still being clarified (Vandekerkhove and Cardoso 2010, Wittfoth and Zettler 2013).

The first limitation, on the other hand, is a critical area for the development of eco-hydraulics and constitutes the central question addressed in this work: contributing to methods for assessing hydromorphological quality that 1) describe the physical structure of the fluvial environment in a more organism-centered fashion, 2) address causality and 3) account for natural variability.

The general principles outlined in chapter 5 have been derived from existing concepts and approaches in the ecological literature, and provide a useful starting point in this task.

Table 4.6: Summary of variables for hydromorphological assessment employed in LAWA (2000) (own translation of the original German version).

Uses: None, navigation, hydropower, flood protection, urbanization
Planform: Curvature, longitudinal sediment banks, bend erosion, other structures (driftwood jams, fallen trees, islands)
Longitudinal profile: Artificial structures (sills, ramps, drop structures, barrages), level of impoundment, canalization, transversal sediment banks, current diversity, depth variability
Cross-section: Type (natural, near-natural, erosional, trapezoidal), depth, bank erosion, width variability, culverts
Bed structure: Substrate, bed revetment, substrate diversity, special structures (riffles, rapids, pools, dead wood, backflow, roots, macrophytes, cascades)
Bank structure: Indigenous vegetation (forest, reeds, shrubs, grasses), non-indigenous vegetation, no vegetation (revetment, erosion), bank protection (bio-engineering, rip-rap, bankface grasses, revetment, wall)
Surrounding area: Land use (forest, reeds, fallow land, crop, park, urban), riparian zone (extensive forest/succession, riparian belt, bordering vegetation), other surrounding structures (waste disposal site, excavation, roads, flood protection structures)

4.2.4 Changes in the scientific view of rivers

Scientific views of river systems have changed over the last decades as the result of inputs from fields such as limnology, lotic community ecology, geomorphology, and landscape ecology. In attendance to this, developing biological criteria for hydromorphological quality requires adopting a perception of river systems that is in accord with the current state of knowledge.

Early views of rivers as channels in which longitudinal gradients represent the main source of variation (Newbold et al. 1981, Vannote et al. 1980, Ward and Stanford 1983) have been modified, with alternative perspectives being proposed in order to account for relevant lateral (Junk et al. 1989, Tockner et al. 2000, Ward and Stanford 1995a,b) vertical (Stanford and Ward 1993) and temporal (Poff et al. 1997, Stanford et al. 2005, Townsend 1989) processes.

At the same time, views of rivers as landscapes, i.e., as internally heterogeneous entities within a heterogeneous context, have given rise to the field of riverine landscape ecology (Ward et al. 2002, Wiens 2002). From this perspective, river systems are studied as a dynamic mosaic of resource patches of different quality organized in a hierarchy of spatial and temporal scales (Poole 2002).

These conceptual developments have helped understand rivers as more than a channel with or without an adjacent floodplain, an arguably distorted notion resulting from the long-term stabilization of their natural variability in much of the world (Ward et al. 2002).

Though not fully, some elements of this new perspective do permeate the general view of river systems in the current framework. For instance, despite the fact that floodplains are not mentioned explicitly in the WFD, CIS Guidance Document No. 12 (EC 2003) does contain provisions concerning their inclusion as water bodies in the planning process.

Similarly, European Standard 14614 (CEN 2004) addresses floodplain functionality explicitly by means of assessment categories related to channel-adjacent land-use, channel-floodplain connectivity and lateral channel movement.

Lastly, although strongly concentrated on main channel features, the River Habitat Survey method (Raven et al. 1998) and the German field mapping method (LAWA 2000) also account for some aspects of floodplains (up to 100 m from banktop), such as intensity and extent of channel-adjacent land use.

However, strategies still need to be developed in order to fully incorporate modern views of rivers into hydromorphological quality assessment.

In this vein, the methodological framework proposed in this thesis aims to *quantify* hydromorphology in a way that 1) goes beyond a static description of structural features and 2) is grounded on ecological knowledge concerning habitat-biota relations for lotic populations.

4.3 HYDROMORPHOLOGY-MACROZOOBENTHOS RELATIONS

Assessing hydromorphology from a biological perspective implies that quality should be measured according to what is favorable to the biological populations of interest. According to the framework proposed in Schuwirth (2012), the factors defining the environment of lotic macrozoobenthos, and therefore also defining what favorable conditions are for these populations, can be organized in five factor groups: **water quality**, **local physical habitat**, **food sources**, **biotic interactions**, and **colonization and spatial processes** (table 5.1).

This factor-group perspective is adopted in the conceptual basis of this thesis in order to clarify the role of hydromorphology as an abiotic variable in macrozoobenthos ecology (see section 5.1). It is used here simply in order to organize the following description of the state of the art on this topic.

4.3.1 Water quality

The relationship between hydromorphology and water quality is, in general, indirect. The major causes of water quality deterioration in lotic systems are point and diffuse pollution sources, with inputs from agriculture and hazardous substances playing a central role in industrialized regions (e.g., in Europe; EEA 2011, 2012a).

Hydromorphology's influence on water quality, in contrast, is most significant in impounded reaches (Kirchesch et al. 2006) and floodplain water bodies with reduced connectivity (Tockner et al. 1999), where hydraulic and hydrological changes can affect the mass balance of different substances (e.g., seston, dissolved, oxygen, sediments, nutrients).

This does not imply that morphology and hydraulics cannot play an important role in determining the chemical composition of water in streams. Small-scale differences in turbulent mixing can lead to heterogeneity in the distribution of dissolved oxygen (Lancaster et al. 2009) and suspended materials (Hart and Finelli 1999), which are relevant to benthic organisms. Similar effects can also exist at higher scales (Vannote et al. 1980).

However, these changes are usually part of the natural dynamics of running water ecosystems, and only high-impact hydromorphological interventions, such as large impoundments or floodplain cut-off, can bring about significant changes in water chemistry. Further, improvements in the chemical status of surface waters will come largely from the management of point and diffuse sources (EEA 2012a), with contributions due to improvements in hydromorphology being rather marginal in most cases.

In sum, given that 1) hydromorphology is not the main factor acting on water quality, 2) water quality analyses can be approached using existing methods, including spatially explicit simulation tools, and 3) the focus of this work is exclusively on hydro- and morphodynamics, water quality considerations are not included.

4.3.2 Biotic interactions

Flow has been found to mediate the intensity of competition, predator-prey interactions and grazing in stream invertebrates, with mechanisms being related to the influence of roughness layer hydrodynamics (Nikora 2010) on the distribution of limiting resources and encounter rates between individual organisms (Hart and Finelli 1999).

Next to hydrodynamic effects, which are due to the mechanical action of water motion, flow can influence biotic interactions through its effects on the temporal distribution of resources. Species with life history strategies (e.g., oviposition, emergence) that are synchronized with the occurrence of specific environmental conditions can be affected by changes in the timing, frequency and predictability of flow events (Poff et al. 1997).

Thus, flow regime alterations can benefit generalists by shifting the habitat template away from the requirements of stenotopic taxa (Poff and Ward 1990). The competitive success of invasive species exemplifies this to some extent, if we consider that the widespread alteration of natural flow regimes (Nilsson et al. 2005) has created environmental conditions that favor exotics and affect native species (Bunn and Arthington 2002).

Certain aspects of the linkage between flow, morphology and biotic interactions are approached explicitly in 'hydraulic food-chain models' (Power et al. 1995). These models relate hydromorphological characteristics, such as depth, width, velocity and floodplain isolation, to rate parameters in Lotka-Volterra-type equations (table 4.7) describing the mass balance for different trophic components (e.g., D'Angelo et al. 1997, Doyle 2006). An example of these relationships is given in table 4.9.

However, these relationships are strictly phenomenological descriptions, and thus constitute a strong simplification (as discussed in Power et al. 1995) of the intricate interactions between hydromorphology and biological populations.

In fact, the study of these dynamic relationships, which are a necessary complement to the mass balance equations used in the aforementioned models, will likely be at the center of eco-hydraulics' development in the coming years. Models featuring these mechanistic relations, described by some authors as 'process-based environmental response models' (Lancaster and Downes 2010a), represent a central short-term goal of eco-hydraulic research.

The relationship between hydromorphology and biotic interactions will not be included in the approach presented in this thesis. The reasons for this are summarized at the end of the sub-section on food sources (section 4.3.3).

4.3.3 Food sources

Hydraulic food-chain models also address the relationship between hydromorphology and food resources to some extent. In the modeling approach presented by D'Angelo et al. (1997), for instance, the biomass of benthic invertebrates is simulated as a function of four environmental drivers (incident radiation, discharge, water temperature and nutrients) via trophic relations similar to those used by Power et al. (1995).

Table 4.7: Biomass balance equations for trophic dynamics subject to hydraulic constraints (modified from Power et al. 1995). For symbols see table 4.8.

Food web component	Mass balance equation
Detritus	$\frac{dD}{dt} = I + m_v V - c_h HD - m_d D$
Vegetation	$\frac{dV}{dt} = rV \frac{K-V}{K} - c_h HV - m_v V$
Herbivore-detritivores (grazers)	$\frac{dH}{dt} = b_h c_h HV + b_h c_h HD - c_p HP - m_h H$
Predators	$\frac{dP}{dt} = b_p c_p HP - m_p P$

Table 4.8: Symbols in table 4.7

Symbol	Variable
D	Detritus standing stock
t	Time
V	Vegetation biomass
H	Grazer biomass
P	Predator biomass
b_h	Conversion efficiency for grazers eating vegetation or detritus
b_p	Conversion efficiency for predators eating grazers
c_h	Per capita grazing rate on detritus or vegetation
c_p	Per capita predation rate on grazers
r	Maximal intrinsic rate of increase for vegetation
I	Input of allochtonous detritus
K	Carrying capacity (asymptotic biomass) for vegetation
m_d	Loss rate of detritus
m_V	Non-grazing mortality of channel vegetation
m_h	Mortality of grazers not due to predation
m_p	Mortality of predators

Channel morphology (e.g., width, roughness) and hydraulics (velocity, depth) are thus assumed to influence macrozoobenthos through their effects on food resources, i.e., algal production, organic matter transport and prey biomass. A similar approach has been used more recently by Schuwirth and Reichert (2013), who based their model (Streambugs 1.0) on a combination of trait information, the Metabolic Theory of Ecology (Brown et al. 2004) and geographic information.

At small scales, hydromorphology can influence the supply of food resources for benthic invertebrates via seston flux (transport and deposition) and flow's mediating role in predator-prey interactions (Hart and Finelli 1999, Lancaster and Downes 2010b). Larger-scale interventions, such as channelization or impoundment, can also influence food resources through their effects on riparian vegetation, either by direct removal or by altering the flow regime.

Along impounded rivers in Sweden, for example, significant changes in species composition, species richness and percentage cover have been found relative to free-flowing reaches (Nilsson et al. 1991, 1997). Similar results have been reported for the Missouri river (U.S.A) (Johnson 2002).

Overall, these changes can influence litter input, which represents the base of trophic pathways involving feeding groups of macrozoobenthos such as shredders, gatherers and filter-feeders (Allan and Castillo 2007).

Other mechanisms can lead to further feeding groups being affected by changes in riparian vegetation. Grazing taxa, which feed on periphytic algae, could also experience alterations in their food supply as a result of increases or decreases in shading following changes in riparian vegetation (Rowell and Sobczak 2008).

Lastly, hydromorphology's relationship with the food supply for macrozoobenthos can also be traced to the interactions between flow, morphodynamics and vegetation distribution in alluvial channels, although these changes take place on higher temporal scales (e.g., decades) (Baptist 2005, Osterkamp and Hupp 2010).

The above discussion illustrates the complexity of hydromorphology's relation to biotic interactions and food resources. Considering these intricate relationships in the methodological framework developed here would require input data for the assessment with a very high level of detail (e.g., see Schuwirth and Reichert 2013).

Simplified criteria, such as the presence/absence of riparian vegetation or its cover, are already used in existing approaches and therefore would not constitute and advance relative to the current state of knowledge. At the same time, crucial trophic pathways such as detritus, seston, periphyton and living prey would be left out of the assessment, making it incomplete and potentially misleading.

For these reasons, these elements are excluded from the general approach developed in this dissertation.

Table 4.9: Description of the relationships between rate parameters in the biomass balance equations (table 4.7) and morphological and hydraulic characteristics (modified from Power et al. 1995).

Parameter	**Response**
c_h	Decreases linearly with velocity after a certain threshold, and ramps down to zero at the slip speed
c_p	Decreases with width, due to higher proportion of refuge area on floodplain. In channel, increases above a threshold velocity at which flow dislodges and washes away refuges
I	Increases with width, due to higher litter input over floodplain
K	Decreases with depth due to light limitation
r	Increases with velocity due to increases in nutrient flux
m_d	Decreases with depth due to temperature or oxygen stratification

4.3.4 Colonization and spatial processes

The ecological success of river restoration measures is ultimately defined by the arrival of organisms (or propagules) to the newly created habitat and the development of viable populations in it. Unless stocking measures are undertaken, this invariably requires dispersal from populations existing elsewhere in the catchment (Kail et al. 2012). Colonization is therefore a critical process in management and restoration under the WFD.

Similar considerations may apply with respect to habitat quality assessment. In dynamic patchy environments, the spatial distribution of resources and habitable space can change over organism-relevant time scales, making local extinction and colonization events critical for population persistence in the landscape (Hanski 1999b).

Given that riverine environments fit this 'patch dynamics view' quite well (Poole 2002, Pringle et al. 1988, Stanford et al. 2005, Townsend 1989), spatial and temporal scale considerations regarding the dynamics of lotic populations should play an important role in human perceptions of habitat quality. In fact, important insight has been gained recently in Germany from this spatial understanding of river systems (e.g., Kail et al. 2012, Schröder et al. 2012, Sondermann et al. 2012).

Habitat quality should therefore be seen as extending beyond the microhabitat scale in attendance to the fact that population persistence, which ultimately determines ecological status and potential under the WFD, is the result of processes that take place at different spatial and temporal scales.

Further, given that the populations of many taxa of macrozoobenthos are likely to be open at the local scale (Bilton et al. 2001, Diehl et al. 2008, Downes and Reich 2008, Lancaster et al. 1996, Palmer et al. 1996), habitat quality definitions should consider scale and spatial processes explicitly.

The role of hydromorphology in this context has been reviewed by Hart and Finelli (1999). The studies cited therein have demonstrated that flow velocity, turbulence and substrate can play a central role in the processes involved in drift, i.e., entrainment, transport and settlement, and its success as dispersal strategy.

Early empirical studies in this respect (e.g., Elliott 1971) determined that the relationship between the number of drifting individuals and transit distance can be approximated by an exponential decay function, and that the decay rate tends to show an inverse relationship with flow velocity (Lancaster et al. 1996).

Local morphology has also been found to influence the rate of return to the stream bed, mainly through effects on the spatiotemporal distribution of turbulence, depth and dead water zones (Bond et al. 2000, Lancaster et al. 1996).

Through its effects on drift dispersal, hydromorphology is likely to play an important role in the spatiotemporal distribution of lotic macrozoobenthos in specific situations. This, however, may vary from taxon to taxon and location to location (Downes and Reich 2008).

In any case, considerations concerning spatial dynamics help expand the current perspective, which is based on the tacit assumptions that 1) ubiquitous drift tends to distribute populations over long stretches of channel and 2) adaptive preferences for sites featuring particular physical characteristics drive the distribution of macrozoobenthos in the river landscape (Downes and Keough 1998, Downes and Reich 2008, Palmer et al. 1996).

Apart from drift, which is an in-stream process, many macrozoobenthos taxa are capable of overland dispersal. This can take place either actively, as flying adults (as for many groups of insects), or passively in the form of vector-transported dormant stages or propagules (Bilton et al. 2001).

Unfortunately, the effects of the characteristics of the stream corridor (e.g., morphology, riparian vegetation) on overland dispersal are still poorly known, and data on dispersal distance, frequency and direction remain scarce for a majority of species (Bilton et al. 2001, Sundermann et al. 2011). Available studies, however, show that these interactions are strongly taxon-specific (Macneale et al. 2005, Petersen et al. 2004).

Based on the above discussion, the methodological framework proposed here recognizes the importance for population persistence of the spatiotemporal dynamics of habitat. Thus, the description of hydromorphology presented here is centered around this notion. At the same time, however, it is also acknowledged that much research is still needed concerning dispersal rates for many taxa, particularly considering the high biological diversity of macrozoobenthos

4.3.5 Physical habitat

By definition, the most direct relationship of hydromorphology with macrozoobenthos is given through physical habitat. This has been the focus of a significant proportion of eco-hydraulic research so far, and is the aspect most extensively dealt with in existing approaches, such as habitat modeling. Taxon-habitat relationships are at the core of this area, particularly those describing microhabitat associations in terms of current and substrate preference.

As defined by Schuwirth (2012), physical habitat involves factors such as local temperature regime, hydrodynamic environment and substrate. The biological suitability of these conditions is commonly expressed in the form of indices with different types of coding, with values being assigned in such a way as to reflect each taxon's affinity for a specific substrate, velocity or temperature range (Leclerc 2005).

This may be done using binary (presence/absence, 1/0) or ordinal (0 - 10) classifications (e.g., Schmidt-Kloiber and Hering 2012). As described in section 5.5, this information is derived from long-term and extensive observations of the distribution of each taxon in the different microhabitats of river landscapes.

This type of numerical coding is widely applied in simulation approaches used to describe the role of physical habitat. Schuwirth and Reichert (2013), for instance, account for habitat conditions in their model by including a 'self-inhibition term' (K_{dens}), which is expected to reflect 'habitat capacity' as defined by the microhabitat preferences of each taxon. This term is the result of multiplying the 'half-saturation density of biomass' under optimal habitat conditions (h_{dens}) with microhabitat suitability (a number between 0 and 1) relative to current ($f_{current}$), temperature (f_{temp}) and substrate ($f_{substrate}$):

$$K_{dens} = h_{dens} \times f_{current} \times f_{temp} \times f_{substrate} \qquad (4.1)$$

Other models also make use of this multiplicative approach, in which each individual factor has the potential to act as limiting condition (Leclerc 2005). Crucially, however, Schuwirth and Reichert's (2013) model is much more comprehensive ecologically, physical habitat quality being only one of the many components used in the prediction of community composition and biomass. This constitutes a fundamental epistemological distinction between this approach and traditional physical habitat modeling, as only the former produces predictions that can be tested against adequate observations.

Thus, for many taxa of lotic macrozoobenthos, habitat relations with respect to local physical habitat are comparatively well known (e.g., Schmidt-Kloiber and Hering 2012), and this knowledge allows assessment on a taxon-specific basis for many groups. Therefore, habitat relations are used as starting point in this thesis. Although it is considered by its authors as part of physical habitat (Poff and Ward 1990, Schuwirth 2012), temperature's effects are excluded here due to the strictly hydraulic-morphological focus of the thesis (see section 6.1).

Apart from mineral substrates (sediment of different sizes, from clay to boulders), the physical structure of macrozoobenthos habitat also includes vegetation and woody debris. The structural role played by these organic substrates in the life of many taxa is widely recognized in stream ecology, and it is considered explicitly in widely used substrate classification schemes (ÖNORM-M-6232 1997).

Next to direct consumption (xylophagy) and feeding substrate (consumption of epixylic biofilm), woody debris can be used for refuge and as a stable attachment, emergence or spawning substrate, particularly in the absence of stony material (comprehensive reviews in Hoffmann and Hering 2000, Pitt and Batzer 2011).

Patches of woody debris also contribute to macroinvertebrate habitat quality by creating obstacles and dead water zones that can accumulate organic matter (Daniels 2006), thus providing food resources under adequate current and substrate conditions for many taxa.

As the result of these interactions, a positive association can exist between different taxa of macrozoobenthos and woody debris (Schneider and Winemiller 2008). Consequently, many restoration projects use this type of material as a way of providing adequate substrate under different circumstances, e.g., in highly degraded urban channels (Vosswinkel et al. 2013) and German waterways (Kleinwächter and Schilling 2013).

Despite its potential benefits, it is important to mention that the introduction of woody debris should always be done in the context of systemic restoration approaches (Beechie et al. 2010). Otherwise, structural benefits could be overshadowed by unintended negative effects on other system components, such as water quality or food resources (Leal 2012).

The structural role of live vegetation has also been demonstrated in different studies. Like woody debris, aquatic plants constitute a structurally complex substrate that can be used for shelter against predators and flow, and as a direct (phytophagy) or indirect (periphyton) food source (Cremona et al. 2008, Gerrish and Bristow 1979, Humphries 1996).

In turn, riparian vegetation can exert an important control on the physical characteristics (structure, temperature, illumination) and resource supply (food sources, woody debris) of both instream and out-of-stream habitats (Moore et al. 2005), thus constituting another important component of the physical habitat of lotic environments.

In view of this, the structural role of live vegetation and woody debris is included in the definition and analysis of hydromorphological quality in this thesis.

5 CONCEPTUAL BASIS

The following conceptual basis derives from a (non-exhaustive) review of the ecological literature on river hydromorphology, habitat-biota relationships in streams, stream assessment and restoration and ecohydraulics. The objective of this review was to gather ideas from recent conceptual and methodological developments in these areas that can be applied to the biological analysis of hydromorphology.

The outcome of this review is presented as a synthesis of these ideas in connection with the derivation of the methodological framework. Principles and concepts from the reviewed literature were selected on the basis of their applicability to the problem at hand and the extent to which they can help further introduce modern ecological thinking into eco-hydraulics and hydromorphological analyses.

5.1 POTENTIAL VS. ACTUAL BIOLOGICAL EFFECTS

In the context of the WFD, the ecological status/potential of running waters with respect to benthic macroinvertebrates is determined by community-level indicators such as taxa richness, taxonomic composition, abundance, and functional composition (Meier et al. 2006). This means that the environmental factors that influence the populations[1] making up the target community are important under the WFD.

From an ecological perspective, these factors can be grouped in numerous ways:

- in the WFD, the quality of river systems is thought of as being made up of physicochemical, hydromorphological and biological elements. This model is very general, and is meant to allow the WFD to encompass a large number of water body types, both surface and sub-surface. However, this quality-oriented perspective, despite being useful for status assessment, is less adequate for addressing causal relationships, which is what is needed in restoration and impact assessment;
- from a population dynamics perspective, populations are affected by endogenous (density-dependent) and exogenous (density-independent) factors (Turchin 2013);

[1] It is important to bear in mind that what is commonly sampled in macrozoobenthos surveys is a complex mixture of many populations, including many groups with terrestrial phases and a still largely unknown spatial population structure (Lancaster and Downes 2010b). Hence, the word 'population' here is used only to refer to this level of biological organization, not to the 'classical', spatially-closed population of early models.

- in landscape analyses (Kuhn and Kleyer 1999), factors affecting populations at this scale can be grouped into environmental factors (e.g., pH, temperature, light exposure), consumable resources (food, shelter, migratory pathways) and variability (monthly, seasonal, annual, disturbance regime);
- from the point of view of the spatial structure of populations, the environment can be understood as a dynamic mosaic of patches offering 'habitable living space' and/or food resources (Downes and Reich 2008) interspersed in a matrix of non-habitat.

Recently published methods in stream macrozoobenthos ecology (e.g., Schuwirth 2012) view the presence or absence of a taxon at a site as the result of five fundamental factor groups (see table 5.1 and figure 5.1).

In this context, hydromorphology is seen as one of the five factor groups influencing macrozoobenthos. This analytical framework allows understanding the relationship between hydromorphology and a site's macrozoobenthos as only one piece of a larger puzzle, in which all factor groups contribute to these organisms' environment as a whole (figure 5.1).

Hydromorphology can thus be understood as a limiting factor (Lancaster and Belyea 2006) rather than as a predictor in stream community ecology, given that prediction requires all relevant factor

Table 5.1: Factor groups influencing the occurrence of a macrozoobenthos taxon at a site (Schuwirth 2012).

Factor group	Description
1	Water quality: pH, oxygen, acidity, salinity, pollutants
2	Biotic interactions: competitors, mutualists, parasites, predators
3	Existence of food sources: leaf litter, organic matter, other macroinvertebrates, algae
4	Spatial processes and colonization: availability of source populations, dispersal and connectivity
5	Local physical habitat: temperature, **hydromorphology (morphology and hydrodynamics)**

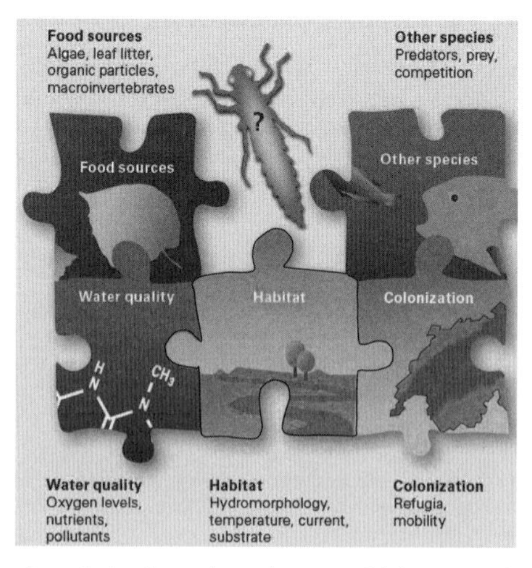

Figure 5.1: Factor complexes that make up the environment of lotic macrozoobenthos, represented as pieces of a puzzle (Schuwirth 2012).

groups to be included in the analysis. Hence, any attempt to link hydromorphology to biological responses must necessarily focus on *potential* rather than *actual* effects.

The use of tolerance ranges for mapping habitat also forces us to look at hydromorphology as determining a potential rather than as a central-type predictor. The reason is that biological responses at the population level result from thresholds in the tolerance of individuals being exceeded by environmental stress (Slobodkin 1968).

Hence, responses at the population-level are by nature threshold-type responses that cannot be analyzed using central-response models. A habitat map therefore indicates the areas where tolerance ranges would not be exceeded if they were inhabited by the target species, i.e., they reflect a habitat potential.

This interpretation of habitat quality is further supported by the results of several field studies on the relationship between macrozoobenthos and habitat in stream ecosystems (Downes et al. 2000, Downes and Lancaster 2010, Fonseca and Hart 2001, Heino et al. 2004, Sharpe and Downes 2006).

According to this body of evidence, the 'repeated finding that benthic densities of channel units, such as riffles, can be as different as samples collected from different catchments or tributaries, even when the riffles are in close spatial proximity' (Downes and Reich 2008), is a strong indication that physical habitat cannot be assumed to drive density and distribution through mechanisms involving habitat preference.

In sum, the evidence strongly suggests that macrozoobenthos distributions are not the result of optimal, central-type responses to physical habitat. Rather, physical habitat provides a template on which many other (including stochastic) factors exert their influence on populations to determine observed distribution and density patterns within this maximum habitat potential.

This distinction between potential and actual effects is one of the main organizing principles of the proposed method. It allows clarifying the possibilities and limitations of any approach for judging stream hydromorphology from an ecological perspective. In particular, it implies that the indicators proposed in the development of the framework will be treated as limiting factors rather than as central-response predictors (Lancaster and Belyea 2006) (figure 5.2) in the interpretation of their effects.

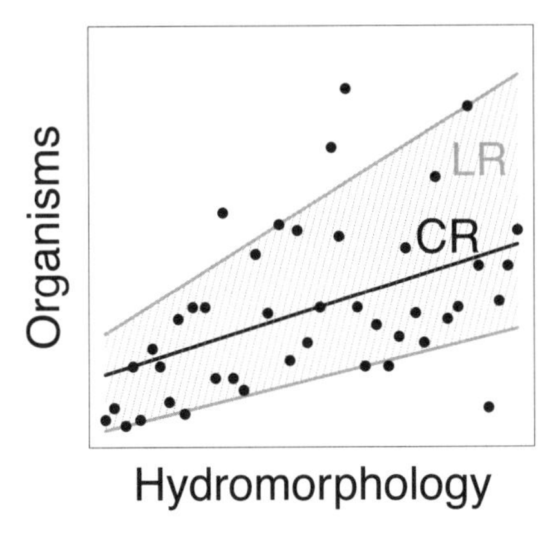

Figure 5.2: Conceptual representation of the difference between central response (CR) and limiting response (LR) approaches (Lancaster and Belyea 2006) using hypothetical data.

This alternative treatment of the indicators implies that they are not judged by their explanatory power (e.g., % variance explained), as is normally done in central response models (using the terminology of Lancaster and Belyea 2006). Rather, the analysis concentrates on whether and how the limits of the distribution of the dependent variable vary along the value of the indicator.

In the hypothetical example of figure 5.2, hydromorphological quality leads to uncertain predictions about the value of the biological variable in the y axis. However, it clearly plays a role in determining the position of the limits within which this variable is observed to vary.

In ecological terms, this allows acknowledging the influence of the other (unmeasured) factor groups instead of referring to it as 'statistical noise', while at the same time providing a clear picture of the influence of hydromorphology. Statistical techniques such as quantile regression (Koenker and Bassett Jr. 1978) provide the mathematical framework for this analyses.

The role of factor groups 1, 2 and 3 in table 5.1 has been addressed explicitly or implicitly in some of the existing approaches to habitat (e.g., CaSiMiR-Benthos [Kopecki and Schneider 2010], INFORM-MOBER [Giebel et al. 2011]) and water quality modeling available nowadays. In contrast, spatial processes have only recently been brought into consideration in river bio-assessment and restoration (e.g., Kail et al. 2012, Schuwirth and Reichert 2013).

For these reasons, the methodological framework presented here is meant to describe and assess habitat **quality** for specific taxa rather than predict their occurrence, abundance or biomass.

Specifically, the framework focuses on what Downes and Reich (2008) call 'habitable living space', i.e., areas of habitat whose hydrodynamics and substrate are considered adequate given the known habitat associations of the target taxa. In order to highlight the fact that the method focuses on in-stream habitat, the term **aquatic habitable space** is used (see section 6.1).

5.2 SCALE AND MACROZOOBENTHOS HABITAT IN STREAMS

Scale in ecology refers to the resolution (or grain) and extent (or range) of observation, which applies to both the temporal and spatial dimensions in which patterns and processes are studied (Turner et al. 2001).

A large body of literature has developed in the last four decades concerning scale in the study of ecological processes. Scale is perhaps the most ubiquitous topic in the field of landscape ecology (Turner et al. 2001); modern lotic community ecology (Leibold et al. 2004, Poff 1997, Townsend 1989), species distribution modeling (Latimer et al. 2006, Liebhold and Gurevitch 2002) and metapopulation ecology (Hanski 1999b) all deal, implicitly or explicitly, with scale in their conceptual and quantitative models.

In river ecology, these developments have led to current views of stream ecosystems as hierarchical systems (Frissell et al. 1986) (figure 5.3), as well as to the emergence of the field of 'river landscape ecology' (Wiens 2002).

These views result from the application of hierarchy theory in ecology (O'Neill 1986), whereby it is recognized that existing large-scale features and gradients, such as regional geology, relief and climate, impose constraints on patterns and processes at intermediate and small spatial and temporal scales.

In turn, phenomena at the lower end of the scale hierarchy, such as turbulent variations, microtopography and individual-level responses, provide mechanistic explanations for the patterns observed at higher levels. In other words, patterns observed at intermediate or higher scales can be understood as the integrative (*sensu* Orians 1980) outcome of the underlying 'low-level dynamics'.

This scale dependency is simply the result of gradients at one scale being stronger than those at other scales. At large observation scales, geological, biogeographic and climatic gradients override the influence of localized factors. In contrast, at small observation scales, most of the variation is explained by small-scale gradients affecting individual organisms' physiology and behavior.

An application of these principles relevant for the purposes of this thesis is illustrated by the study of Downes et al. (2000). These authors studied variability in stream macroinvertebrate density and diversity using a nested sampling design that involved substrate types within sites (micro-scale), sites within reaches (meso-scale), and reaches in three rivers (macro-scale) of the same catchment.

Substrate-independent inter-site differences explained approximately 30% of variation in species richness and 60% of that in density of individuals, showing that meso-scale variability can override local (substrate) and large-scale (river) effects in stream macrozoobenthos communities.

A series of similar, independent studies (Beisel et al. 2000, Boyero 2003, Downes and Reich 2008, Downes and Lancaster 2010, Fonseca and Hart 2001, Heino et al. 2004, Sharpe and Downes 2006, Thompson and Townsend 2006, Townsend and Hildrew 1976) have found results consistent with these, which can be seen as a corroboration that stream macrozoobenthos populations have a strongly patchy distribution (Minshall and Minshall 1977). At the same time, these findings indicate that within-reach variability at scales between a few tens of centimeters and a few meters plays an important role in the long term persistence of these organisms.

These findings are in line with concepts in modern stream ecology such as that of 'meta-structure' (Poole 2002), according to which the size, distribution and temporal variability of habitat within a reach can be seen as a major determinant of the behavior of the reach as a whole, particularly in highly dynamic habitats such as streams.

For the relationship between hydromorphology and macrozoobenthos, this implies that the former's role as a habitat component should be described by looking at the internal spatiotemporal heterogeneity of the reach rather than by building reach-scale aggregate measures. Further, aggregating hydromorphology over an entire reach, as is done in existing assessment methods,

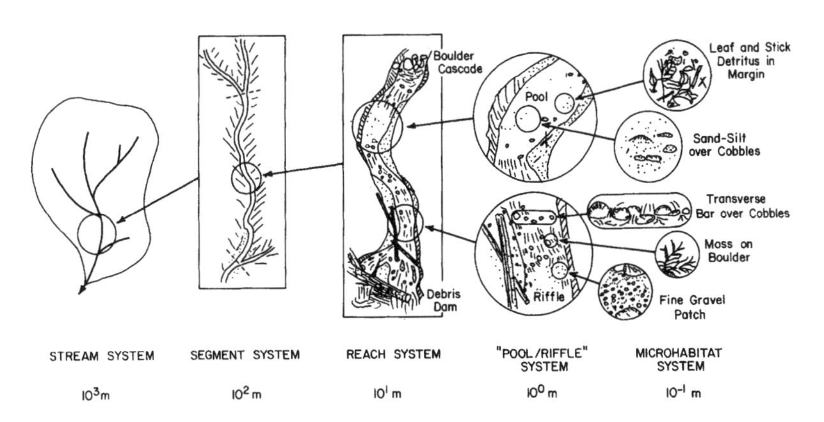

Figure 5.3: Hierarchical organization of stream systems and their habitats proposed by Frissell et al. (1986).

averages out variations in substrate, flow velocity and water depth that, according to the findings described above, are relevant for the persistence of stream macrozoobenthos.

From this discussion, it follows that the scales for describing stream habitat must be selected based on what is relevant for the survival of the target organisms, which is the central tenet of the concept of 'ecological neighborhoods' proposed by Addicott et al. (1987).

This concept states that observation scales in the analysis of habitat-biota relationships must be determined based on the space and time in which the target ecological process takes place. For hydromorphology-macrozoobenthos relations, the target ecological process is the long-term survival of the target species in the studied reach, which, as discussed above, is linked to within-reach variability (meta-structure) at scales between a few tens of centimeters and a few meters.

Based on these principles, this dissertation approaches the analysis of hydromorphological quality by exploring within-reach variations in hydromorphological quantities in time and space. A detailed description of the variables and scales of the approach is presented in section 6.2.

5.3 TEMPORAL VARIABILITY AND DISTURBANCE

Ecological patterns and processes are variable in time, and this variability occurs at different temporal scales. For instance, it has been known for more than two decades that re-colonization of newly created habitat can take longer in catchments with isolated source populations than at sites surrounded by good-quality, well-populated habitat (Fuchs and Statzner 1990).

Different processes may dominate at different temporal resolution (frequency) and extent (duration) (Turner et al. 2001). Over short time scales, for instance days, habitat availability may show little or no changes. Increasing extent to several months or a few years will reveal seasonal changes in resources and habitability, and over periods of decades, long-term variations could emerge as the result of long-term environmental and population-level trends. Changes in resolution can lead to similar conclusions; shortening time between samples, for instance, can reveal the effects of fast processes undetectable at lower observation frequencies.

As for space, the selection of temporal scales for analysis should follow what is known about the phenomenon of interest (Addicott et al. 1987, Wiens 1989). In the method proposed here, this implies looking at hydromorphological quality at time scales that are relevant for the long-term persistence of the target species in the studied stream reach.

Therefore, the indicators proposed in this thesis (section 6.2) are computed from dynamic habitat simulations at a resolution and extent that allow characterizing the habitat variability framework to which the target species is exposed. The key parameter in this analysis is the generation time of the target species (Poff and Ward 1990) (see section 6.2.1).

This approach allows accounting for the effects of the hydrological dynamics of the stream in the analysis of hydromorphological quality, which opens new possibilities for characterizing a given morphological configuration as ecologically suitable or unsuitable depending on how these structural features interact with hydrodynamics to determine the spatiotemporal distribution of habitat.

At the same time, extending the analysis over several generations of the target species allows including the potential effects of disturbances[2] on habitat quality (e.g., Poff and Ward 1990, Townsend et al. 1997).

[2] Here, disturbance does not refer exclusively to destructive human intervention, but rather to a 'discrete event in time' (Townsend 1989) of enough magnitude to produce significant changes in habitat and biological communities. Floods or droughts, either natural or man-made, are examples of disturbances in lotic ecosystems as they are understood in this thesis.

From this perspective, lotic macrozoobenthos at a given point in time is not necessarily the result of concurrent environmental conditions. Just as catchment-wide or reach-scale factors can add to, or even overwhelm local effects (e.g., local habitat quality), past events can lead to time lags in biological responses (Gerisch et al. 2012) and even confound the effects of local habitat associations on the distribution of many taxa (Lancaster and Downes 2010a,b).

Recent discussions on river restoration recognize habitat temporal dynamics as a factor that can enhance biodiversity over management-relevant time scales (Beechie et al. 2010). 'Static' or 'stable' habitats, as envisioned in many morphological restoration projects, provide a fixed set of conditions that may well satisfy the requirements of some taxa, but lack the variability that underlies the natural functioning of lotic environments (Palmer et al. 2010, Poff and Ward 1990, Townsend 1989).

For example, variation in time in the distribution of resources and habitable space can allow coexistence of species that would otherwise be excluded by competition or predation, which can happen in habitats with reduced variability (Townsend 1989).

Furthermore, natural environments do not provide optimal conditions for all taxa at all times, and therefore this should not be the goal of management strategies (Anderson et al. 2006). Hence, our view of hydromorphology under the WFD should include habitat frameworks that allow for enough spatial and temporal dynamics.

In running waters, this temporal variability is underpinned in most, if not all, of its aspects by the flow regime. Frequency, magnitude, duration, predictability and timing of flow events determine the characteristics of a great number of processes in river systems (Poff et al. 1997).

Based on the above discussion, the historic time course of discharge is used in this thesis to describe the habitat variability framework of the studied reach. This is achieved though use of a discharge time series as boundary condition in state-of-the-art shallow water hydrodynamic numerical models, which allow translating these discharge variations into a distribution of habitat based on existing knowledge on the habitat associations of the target species (section 6.1).

5.4 THE PHYSICAL HABITAT TEMPLATE AND PATCH DYNAMICS IN STREAMS

Given the importance of spatial and temporal heterogeneity, hydromorphology's role is understood in this thesis as part of what is known in stream ecology as the 'physical habitat template' (Poff and Ward 1990). This concept derives directly from habitat templet theory (Southwood 1977), and emphasizes the role of substrate, streamflow and, most importantly, their **spatiotemporal variability**, as environmental forces acting on lotic populations.

Habitat templet theory (Southwood 1977) is a model in population-evolutionary ecology (Korfiatis and Stamou 1999) that emphasizes the role played by spatiotemporal environmental variability in the evolution of organisms' life history strategies. This includes traits such as dispersal capacity, dormancy, desiccation resistance, body size and the timing of life cycle events, among others.

The principle of temporal habitat heterogeneity outlined in habitat template theory is illustrated in figure 5.4. This diagram presents a hypothetical situation with two species (A and B) featuring different generation times (τ_A and τ_B), both inhabiting an environment that exhibits a sequence of favorable (F_i) and unfavorable (L_i) periods of different duration and 'favourableness' (r).

During favorable periods, $r > 0$ and both breeding and existence are possible for the two species. On the contrary, unfavorable periods ($r \leq 0$) permit existence but not breeding. Thus, a species will be able to persist in the habitat until its generation time τ is shorter than the duration of an unfavorable period. As long as this is not the case, as for species B ($\tau_B > L_i$), the next favorable period will be reached and breeding will again become possible, making persistence more likely.

In contrast, if generation time τ is shorter than L_i, persistence will be less probable. In this case, unsuitable periods cannot be overcome due to the species' shorter generation time ($\tau_A < L_i$). As a consequence, all individuals of species A will die off during L_i before being able to breed again, making the species unlikely to persist in this habitat.

Poff and Ward (1990) apply this conceptual framework to stream ecosystems in order to derive the 'physical habitat template' concept. In this new framework, focus is shifted from evolutionary to ecological time scales with the purpose of explaining the persistence of lotic populations as the result of adjustment to the spatiotemporal variability of the stream's physical habitat.

Thus, these authors move from the development of life history strategies over evolutionary time scales to the persistence of local populations in a stream reach over ecological time scales, which are closer to management time scales in the context of the WFD.

As formulated by its authors, the physical habitat template is 'the long-term regime of natural environmental heterogeneity and disturbance', which is seen as a major determinant of the 'types of species attributes appropriate for local persistence' (Poff and Ward 1990). Hence, it is contended that temporal variability, disturbance and spatial heterogeneity play a critical role in determining the composition of communities inhabiting streams.

This concept is illustrated in figure 5.5, which presents three species, A, B and C (each in a different plot), living in the same hypothetical stream. The curve in each graph indicates the intensity of environmental stress as experienced by each species ('intensity of environmental signal', y axis).

The tolerance range of each species is indicated by the height of the shaded boxes in each plot, whereby higher boxes indicate a higher tolerance level. In turn, the width of each box indicates species phenology, i.e., the time periods in which each generation inhabits the stream. Thus, species A has a generation time of 2 years, and is present in the stream throughout the entire

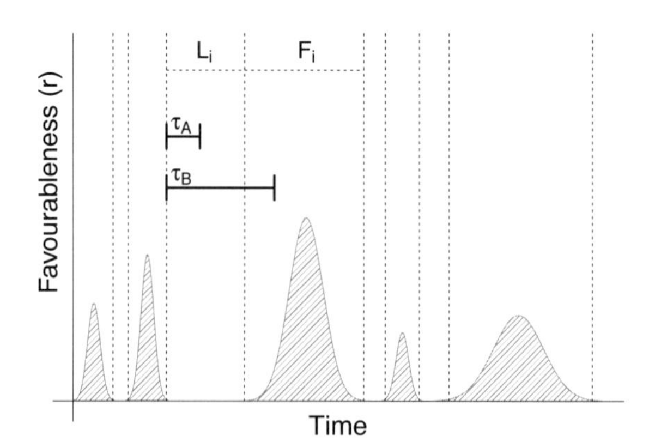

Figure 5.4: Graphical representation of the habitat template concept (Southwood 1977). Temporal variations in favourableness (r) lead to a sequence of favorable (shaded) and unfavorable periods. F_i = length of a favorable period (both existence and breeding are possible), L_i = length of an unfavorable period (only existence is possible). τ_A and τ_B = generation time of species A and B, respectively. Modified from Southwood (1977).

2.5-year period shown in the plot. In contrast, species B and C have generation times of less than a year.

The spaces between generations (gaps between boxes) indicate periods in which only propagules (e.g., eggs), terrestrial adult, or highly resistant dormant stages are present. In these periods, environmental stress plays a less important role in determining population persistence.

Hence, the long-term survival of a species in this hypothetical stream is determined by the intensity of environmental stress during the time periods it inhabits the system. If environmental stress exceeds the tolerance range of the species, individuals will be eliminated from the population, and this will continue until stress falls below the tolerance level.

This is the case for species B and particularly for species C. In contrast, despite always being present in the stream and therefore exposed to environmental stress, species A will not be affected by it due to its high tolerance level. This species' survival will therefore be more likely than for species B and C.

The ecological significance of spatiotemporal heterogeneity in streams is further emphasized by the 'patch dynamics' perspective (Pringle et al. 1988, Townsend 1989). After comparing different models of community organization with empirical evidence from stream systems, Townsend (1989) and Pringle et al. (1988) conclude that a species' survival in a stream is determined by its capacity to use spatially distributed resources (dispersal capacity), its tolerance against environmental fluctuations and, to a lesser degree, its competitive ability.

Figure 5.6 illustrates this concept. The upper left corner represents the extreme case of no spatial and no temporal variability, as in classical Lotka-Volterra models based on the competitive

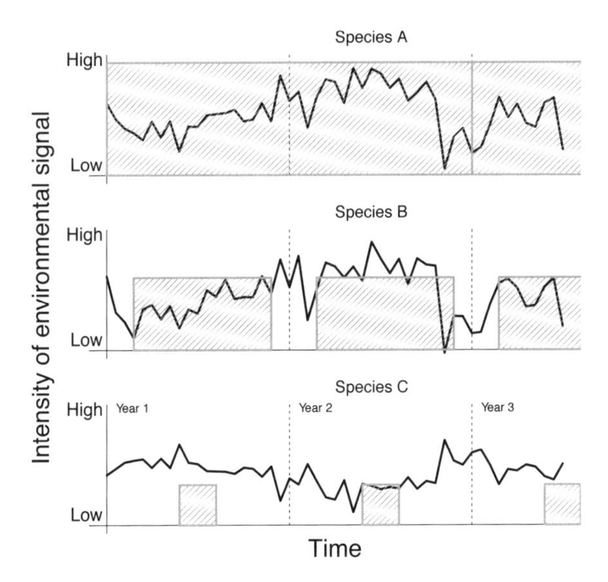

Figure 5.5: Graphical representation of the physical habitat template (Poff and Ward 1990). Each plot corresponds to a different species (A, B and C). Vertical dotted lines represent years, as indicated for species C. Shaded boxes represent each species' tolerance level (vertically) and phenology (horizontally). The curve in each plot represents the intensity of environmental stress (y axis). See text for explanation. Modified from (Poff and Ward 1990).

exclusion principle. In the upper right corner are equally extreme systems in which temporal variability is so great that survival is practically impossible (unoccupied space).

Systems with higher spatial variability but no or little temporal variation (lower left area) lead to communities in which resource partitioning in space determines species co-existence ('niche controlled' communities). As temporal variability increases (to the right), tolerance against environmental fluctuations gains importance relative to competitive ability, leading to a so-called 'successional mosaic' ('dominance-controlled' communities).

Further along the gradient, environmental fluctuation leads to a 'competitive lottery' ('founder-controlled' community), in which the role of competition is even smaller and species persistence is determined by their adjustment to environmental stress. Finally, temporal variability becomes so large as to completely eliminate the influence of competition (lower right corner).

The patch dynamics concept, and implicitly the physical habitat template concept (Poff and Ward 1990), situate stream communities mainly in the 'competitive lottery' region (Townsend 1989) (shaded area in figure 5.6). For management under the WFD, this implies that methods aiming to link hydromorphology and lotic organisms (in this case macrozoobenthos) will benefit from a detailed description of physical habitat's heterogeneity in space and time.

The methodological framework proposed in this thesis adopts this view in order to develop a description of hydromorphology and a set of benchmarks for its judgment. In doing so, it proposes

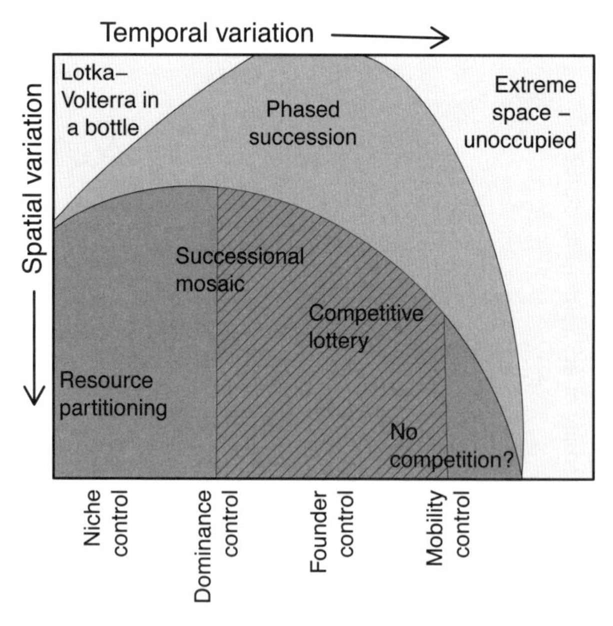

Figure 5.6: Graphical representation of the principle underlying the patch dynamics perspective of streams (Townsend 1989). Different forms of community organization and species persistence are possible under different intensities of spatial (y axis) and temporal (x axis) variability. Darker areas represent greater species richness. The shaded area indicates where stream ecosystems are situated under the patch dynamics concept. Re-drawn from (Townsend 1989).

a strategy that can help establish what should be measured in order to define hydromorphological quality from a biological perspective.

Thus, this thesis presents a view of the WFD-derived notion of hydromorphology that defines it as a component of the physical habitat template of lotic ecosystems. This is useful because

1. The physical habitat template provides a framework for explaining hydromorphology's relation to biological responses that is based on ecological theory,
2. It identifies population persistence under a regime of spatial and temporal variability as the biological process of interest in the ecological analysis of hydromorphology under the WFD, and
3. It allows developing hydromorphological quality indicators with an explicit ecological meaning.

5.5 TAXON-HABITAT RELATIONSHIPS

Existing bio-assessment approaches have produced and organized a vast amount of knowledge on freshwater invertebrates' habitat relations and traits over the last decades (e.g., Moog 2002, Schmidt-Kloiber and Hering 2012). These are so-called taxon-habitat relations or habitat associations.

These relations may include, among many others, preferences with respect to current, substrate and saprobic conditions, as well as morphological and functional traits such as size, general body form, mobility, and reproductive and feeding habits. Though still in development, these databases constitute valuable sources of systematically-organized knowledge.

River bio-assessment methods based on this knowledge can be and are used to evaluate hydromorphological quality (e.g., Giebel et al. 2011, Kopecki and Schneider 2010). The main principle underlying these methods is that knowledge of a species' habitat relations can be used to translate abiotic conditions, such as water depth, flow velocity, substrate type and riparian vegetation, into a habitat quality/suitability value, usually a number between 0 and 1 (Leclerc 2005).

This translation is accomplished by means of preference curves (Stalnaker et al. 1995) or rule sets (Jorde et al. 2001) based on expert knowledge and/or empirical data, mostly observational field studies. Habitat quality can thus be modeled from abiotic data in order to explore different management scenarios.

These models, sometimes called 'habitat association models' (Lancaster and Downes 2010b), draw directly and indirectly on the principles of habitat templet theory (Southwood 1977, 1988), the Hutchinsonian view of ecological niche (Hutchinson 1957) and the levels of biological response to environmental stress proposed by (Slobodkin 1968).

In habitat templet theory, the characteristics of a species (morphological, physiological, and behavioral) are understood as the result of adaptation under the challenges posed by habitat in terms of disturbances and resource level. In turn, Hutchinson's niche views a species' environment as an n-dimensional hyper-volume, with each dimension corresponding to an environmental variable influencing survivorship and the species' tolerance interval for each variable representing the size of the hyper-volume.

Slobodkin's (1968) framework organizes organisms' relation with habitat into three levels of response: behavioral, physiological and population. In the first, unsuitable environmental conditions are avoided by individuals through behavioral mechanisms (movement, hiding, etc.). If stress cannot be avoided, physiological tolerance comes into play, and when these individual-level mechanisms are exceeded, a response takes place at the population level by which individuals are removed from the population.

This last type of biological response integrates (*sensu* Orians 1980) the first two, and it is the most relevant from a management perspective given that ecological status/potential under the WFD is related to the long-term survival of the populations of the directive's biological quality components.

According to these principles, a species' characteristics and its habitat are intimately associated, and therefore knowledge of the tolerance ranges of the species can provide a basis on which to predict its distribution in a given landscape within its biogeographic range (figure 5.7).

These ideas underpin eco-hydraulic approaches to varying degrees, from early methods for physical habitat simulation (e.g., Bovee 1982) to correlation-based habitat association models (e.g., Lamouroux et al. 2010).

Further, trait-based approaches derived from the habitat templet theory have been used in many areas of stream and general ecology, e.g., to study the effect of disturbance and spatial heterogeneity on community organization (Townsend et al. 1997), to quantify the relative importance of multiple stressors (Statzner and Bêche 2010) and to predict the community-level effects of pesticides (Liess and von der Ohe 2005).

The definition of the stream type-specific 'Good Ecological Potential' (GEP) under the WFD also makes use of existing knowledge on taxon-habitat relationships. Due to the lack of reference conditions for Ecological Potential classification (unlike for Ecological Status), the use of simulation approaches has been suggested to provide assessment benchmarks (Bizjak et al. 2006).

In Germany, for instance, estimations of the macrozoobenthos community under GEP conditions are obtained by compiling an aggregate taxa list for the study area, allocating taxa to the habitats where they are typically found, and then using the abundance the different habitats would have

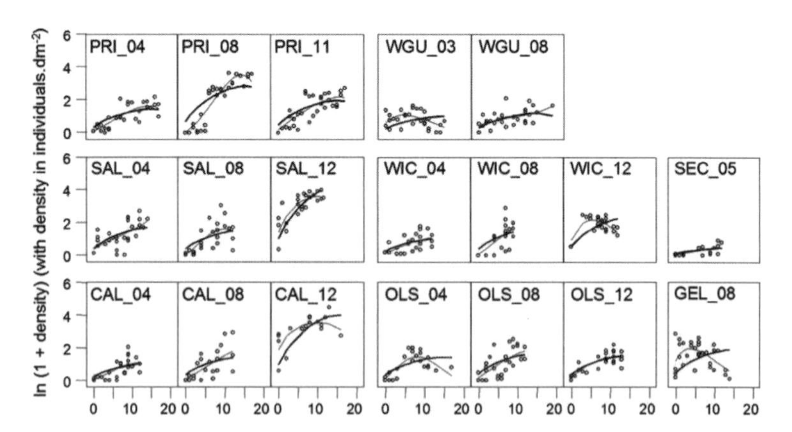

Figure 5.7: Density of mayfly *Baetis rhodani* as a function of bottom shear stress (FST hemisphere number of Statzner and Müller 1989) in Surber samples collected in 19 surveys made in 8 independent German streams at different seasons. The label in each frame indicates the river code and the sampling month. The bold line is an average model fit (common shape in all surveys); the thin line is a stream-specific one. Source: Lamouroux et al. (2010).

under GEP to estimate each taxon's abundance category (Pottgiesser et al. 2008, 2009). The habitat distribution under GEP is estimated using numerical simulation models for flow, morphology and water quality.

Finally, computer-generated random samples of this hypothetical GEP benthic invertebrate biota (Lorenz et al. 2004) are evaluated using the national assessment system (Meier et al. 2006) in order to provide reference values for the biological assessment metrics.

However, like many other methods in ecology, approaches based on habitat associations are not without criticism. Recent reviews of trait-based approaches have identified a series difficulties, mostly related to the non-independence of traits and the fact that their adaptive value is species-specific (Verberk et al. 2013).

In addition, the development of fields such as landscape and spatial ecology (Hanski 1999b, Liebhold and Gurevitch 2002, Turner et al. 2001), together with emerging views of stochastic processes as inherent to ecosystems (Pickett et al. 2007, Shaffer 1981), have accumulated a large body of evidence suggesting that habitat alone is not enough to accurately predict the distribution of organisms in their environment (Latimer et al. 2006). The restricted success of river restoration in central Europe can be seen as an illustration of this.

However, rather than proscribing the use of taxon-habitat relations entirely, the view adopted in this thesis is that their use should focus on the analysis of habitat *potential* rather than prediction of actual standing stocks or distributions.

The methodological framework proposed here acknowledges that habitat 'preference', i.e., active selection of one habitat over others (Downes and Reich 2008, Underwood et al. 2004), is not always the overriding factor determining organisms' distribution, and that chance, dispersal and previous events can also play a significant role (Downes and Lancaster 2010).

At the same time, the framework recognizes that habitat associations are the result of decades of systematic empirical observation, and that they can represent a valid summary of the overall ecological characteristics of many macroinvertebrate taxa (Lamouroux et al. 2010).

Based on this, habitat associations are used in the method for translating hydromorphological information into a distribution of *potential* habitat in space and time, not with the goal of predicting actual biological effects (section 5.1), but rather, as mentioned above, to examine the influence of hydromorphology on the habitat *potential* of the studied stream reach.

This provides further justification for concentrating on potential rather than actual effects in the analysis of hydromorphological quality. This alternative perspective is validated against the actual distribution of macrozoobenthos in section 8.

5.6 TAXON-SPECIFIC AND INTEGRATIVE APPROACHES

The fact that habitat quality is species-specific poses a challenge for stream assessment and restoration. If one were to restore habitat or evaluate its quality on a species-by-species basis for any ecosystem, the task would involve a spectrum of organisms spanning from microbes to trees and mammals, which is clearly impracticable.

Therefore, both restoration endpoints and quality benchmarks need strategies to integrate the system's biological state into a more tractable set of 'aggregate variables' (Orians 1980). Any integration will of course sacrifice information about particular species; however, as long as benefits outweigh losses, this integrative view remains defensible in a management context.

The degree to which integration is possible depends on where the target ecosystem lies along a gradient of internal vs. external controls on local community structure (figure 5.8).

On one end of this gradient are systems in which "variability is very small, i.e., community structure is controlled 'internally' by strong biotic interactions and stochasticity in species membership is reduced" (Palmer et al. 1997). These communities tend to be stable and more predictable in terms of species membership (point A in figure 5.8), and local-regional exchange tends to be small (x axis of the graph on the right in figure 5.8).

On the other end are 'less biologically predictable systems', in which "community 'assembly' may be a function of dynamic dispersal processes and less predictable local interactions among species post-recruitment" (Palmer et al. 1997). Species membership in these communities tends to be a stochastic sample (point B in figure 5.8) of the regional species pool as a result of a stronger local-regional exchange (x axis of the graph on the right in figure 5.8).

In internally/biologically controlled systems (A in figure 5.8), integration should aim to capture community structure, i.e., species composition and relative abundance, their interactions and relations to local habitat (C in 5.8). In contrast, more stochastic communities (B in figure 5.8) should be managed by focusing on species function (D in figure 5.8) rather than identity, since composition may vary randomly in time due to the stronger level of local-regional exchanges.

In practical terms, the above discussion implies that, for internally-controlled systems (A in figure 5.8), biological restoration endpoints and quality standards should be set in terms of taxonomic community structure.

In turn, for dispersal-controlled systems (B in figure 5.8), this should be done using indicators of functional organization, focusing on the role of potential members in key life-sustaining processes, such as organic matter processing, food supply for upper trophic levels, competing alien species, and others (Palmer et al. 1997).

Existing assessment methods (e.g., Meier et al. 2006) include some of the elements of the above discussion. They are based on the idea that a careful selection of metrics related to both taxonomic identity and functional organization can capture the effects of specific stressors (Karr and Chu 1999), a logic also followed by some trait-based approaches (e.g., Statzner and Bêche 2010).

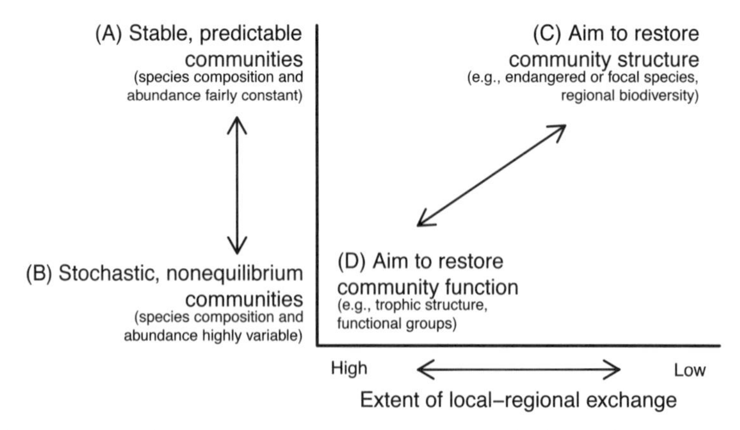

Figure 5.8: Relationship between community stability, extent of local-regional exchange and restoration objectives. Re-drawn from Palmer et al. (1997). For explanation, see text.

However, a full account of the differences between internally- and externally-controlled communities requires allowing for some form of variability in the indicators in the method. In terms of management, this implies a shift from *static* to *dynamic* goals, which is in better agreement with the non-static nature of running waters (Beechie et al. 2010, Palmer et al. 2005, 2010). Evidence that benthic communities as (at least temporarily) open (Downes and Keough 1998, Leibold et al. 2004, Palmer et al. 1996) and spatially dynamic (Downes and Reich 2008, Downes and Lancaster 2010) lend further support to this perspective.

The 'physical habitat template' concept (Poff and Ward 1990) can also be used to address this question. When river systems are viewed as a *dynamic* mosaic of patches of species-specific resources distributed in space and time (Poole 2002, Stanford et al. 2005), it becomes clear that ecological management targets should consider some form of variability.

Further, variability is an integral component of the physical habitat template, and its suitability can be analyzed against the integrated requirements of the target biota (Poff and Ward 1990). Thus, resorting to these two concepts of stream ecology can provide objective biological targets for hydromorphological restoration and quality assessment.

Where stream ecosystems lie along the gradient of community control, and therefore whether taxonomic or functional approaches (or their combination) are more adequate, is still a matter of debate (Allan and Castillo 2007). The objective of the present discussion is solely to use the integrative perspective to address the common question of how to approach hydromorphological quality and restoration given the large number of species that inhabit running waters.

The view adopted in this thesis is that the answer to this question is likely to be strongly case-specific, varying throughout the river system depending on the nature (amplitude, frequency, predictability, etc.) of local dynamics. Whatever the case, though, a shift away from purely deterministic views on community development might be advantageous.

6 DEVELOPMENT OF THE METHODOLOGICAL FRAMEWORK

This dissertation proposes a strategy for assessing the ecological role of hydromorphology based on a combination of ecological concepts and modeling tools available in river hydraulics. The driving principle followed in this process is illustrated in figure 6.1.

In this chapter, a detailed description of the rationale behind this diagram is presented, which shows the combination of three fundamental concepts in ecology into a new interpretation of hydromorphology: the hydromorphological template.

This interpretation is made possible by the use spatially and temporally explicit habitat information derived from numerical shallow water models and substrate data. As is shown in the following sections, this strategy provides a description of hydromorphology that has a more direct link with the stream communities used as indicators of ecological status under the Water Framework Directive.

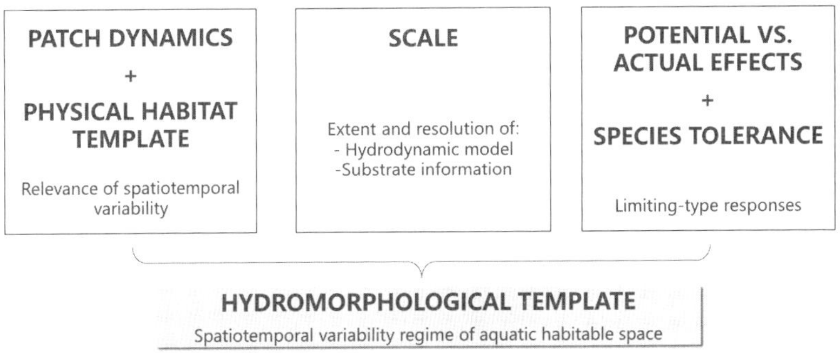

PATCH DYNAMICS
+
PHYSICAL HABITAT TEMPLATE
Relevance of spatiotemporal variability

SCALE
Extent and resolution of:
- Hydrodynamic model
-Substrate information

POTENTIAL VS. ACTUAL EFFECTS
+
SPECIES TOLERANCE
Limiting-type responses

HYDROMORPHOLOGICAL TEMPLATE
Spatiotemporal variability regime of aquatic habitable space

Figure 6.1: Ecological concepts underlying the development of the proposed approach for analyzing hydromorphology from an ecological perspective.

6.1 ELEMENT 1: THE 'HYDROMORPHOLOGICAL TEMPLATE' AS THE DISTRIBUTION OF AQUATIC HABITABLE SPACE

The conceptual basis presented in the last chapter allows re-interpreting hydromorphology in order to make it more suitable for ecological quality analyses. This is achieved by clarifying its role as an abiotic component of the habitat of stream macrozobenthos.

The connection between hydromorphology and stream habitat can be defined as follows. The reach-scale morphology of the stream and its floodplain, together with its substrates and all other hydraulically-relevant features, determine the way water flows in the system, i.e., the hydrodynamics of the reach in terms of the distribution of flow velocities and water depths.

In turn, the overlay of hydrodynamics, substrates, and the tolerance ranges of the target species determines the distribution of hydrmorphologically suitable areas. This process allows excluding surfaces that are physically unsuitable for the target species at a point in time because they are dry, flow is too fast and/or substrate is inadequate.

The result of this overlay is a map of hydromorphologically suitable areas for a given discharge. This interpretation is different from that of traditional physical habitat modeling, in which the resulting area is thought, at least implicitly, to represent habitat in its entirety and is therefore expected to work as a central-type predictor of the distribution of the target species.

However, as discussed in sections 5.1 and 5.4, this is insufficient for drawing conclusions about the long-term survival of the target species and therefore also for judging hydromorphology from an ecological perspective. In short:

- Hydromorphology is only one component of the abiotic environment of stream macrozoobenthos

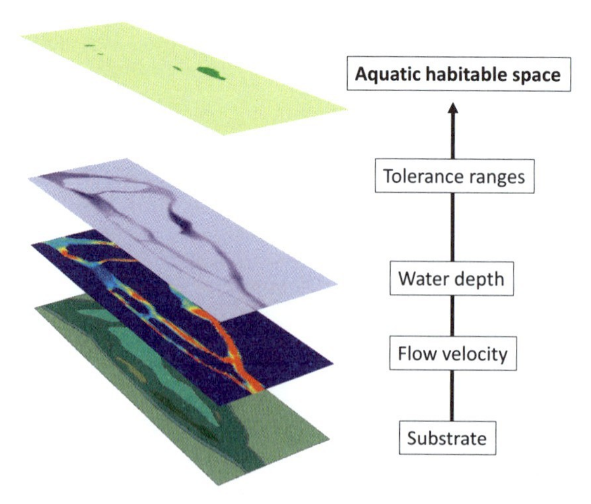

Figure 6.2: The overlay of substrate, water depth, flow velocity and the target species' tolerance ranges allows mapping aquatic habitable space in the study reach.

- Physical habitat suitability represents a maximum potential of habitat, which is modeled using tolerance ranges in attendance to the fact that population-level responses are by nature threshold-type responses

- A large body of evidence supports the conclusion that physical habitat represents a 'stage' on which many other biotic and abiotic factors exert their influence in order to determine the actual size of the target population

From this discussion it follows that a habitat map thus modeled can be interpreted as a map of the surface that is **potentially** habitable in hydromorphological terms. In order to highlight the ecological focus of this definition, the term **aquatic habitable space** (in German = 'aquatischer besiedelbarer Raum'), which stems from aquatic entomology Downes and Reich (2008), is adopted here.

Thus, aquatic habitable space refers to the areas that are suitable for the target species according to their substrate, water depth and flow velocity (figure 6.4). More importantly, based on the principles presented in section 5.1, this space represents a habitat potential and cannot be linked to observed population densities or taxa richness through central-response models.

Nevertheless, if properly described, aquatic habitable space can be used to gain ecologically sound insights about hydromorphological quality by exploring its role as a **limiting factor**. This is achieved by applying the concepts presented in section 5.4 (figure 6.3), which highlight the importance of spatiotemporal variability for the long-term persistence of stream populations.

From this perspective, a map of aquatic habitable space can be seen as a punctual observation in the larger context of the spatiotemporal variability framework of the stream. Consequently, instead of a static view, the ecological analysis of hydromorphology requires a description of the **spatiotemporal variability regime** of aquatic habitable space.

This variability regime results from the contraction, expansion and shifting of habitable space surfaces as they are driven by the hydrological dynamics of the stream. Hence, hydromorphological quality analyses as conceived here are based on the description of this fluctuating supply of aquatic habitable space and the exploration of its limiting effects on the local population of the target species.

Viewing hydromorphology from this perspective allows defining a **hydromorphological template** in order to make its connection with ecological quality more explicit:

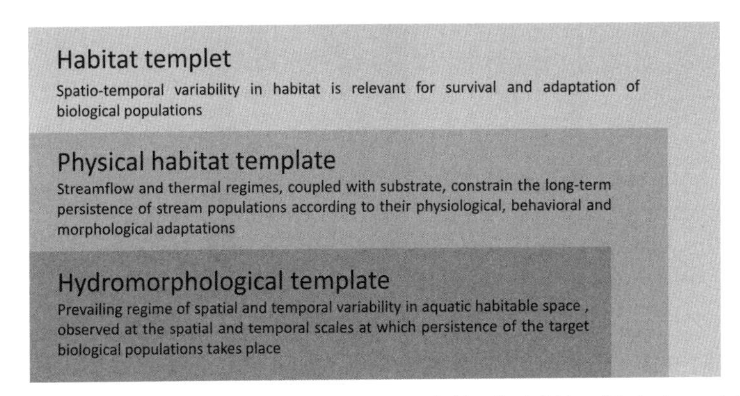

Figure 6.3: Relationship between the theoretical concepts underlying the definition of the hydromorphological template proposed in this thesis.

The hydromorphological template of a stream is the prevailing regime of spatial and temporal variability in aquatic habitable space at the spatial and temporal scales at which persistence of the target biological populations takes place.

Thus, the amplitude, frequency, duration and spatial distribution of aquatic habitable space variations can be used in order to explore the ecological role of hydromorphology. This variability regime can be simulated explicitly in space and time using substrate maps and two-dimensional numerical shallow water models as long as sufficient data are available for model setup, calibration and validation.

However, such a description of the hydromorphological template must take place at spatial and temporal scales that allow capturing habitable space variations that are relevant for the survival of the target species' populations. What survival-relevant variations are in this context and how they can be described at the appropriate scales constitute the central aspect of the second element of the method.

6.2 ELEMENT 2: SCALES AND VARIABLES FOR DESCRIBING THE HYDROMORPHOLOGICAL TEMPLATE

6.2.1 Selecting analysis scales for the hydromorphological template

As discussed in section 5.2, analysis scales in ecology must be set based on the ecological phenomenon of interest, as follows from the 'ecological neighborhoods' concept (Addicott et al. 1987).

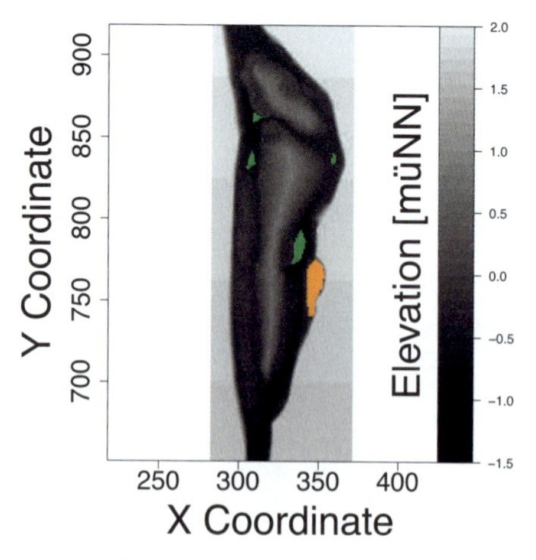

Figure 6.4: Hypothetical example of the supply of aquatic habitable space in a stream reach at two times during the hydrological year. All habitable space surfaces available at t_1 (green surface) disappear at t_2, while a new area becomes available (orange area). The digital elevation model for this hypothetical example was created using data for site cr, see section 8.2.1.

These principles are applied in this section in order to set the spatial and temporal scales at which the hydromorphological template is described in the proposed method. Given that the argumentation for each component of scale is different, the discussion is presented separately for spatial extent, spatial resolution, temporal extent and temporal resolution. The results of this analysis are summarized in table 6.1.

Spatial extent is set as follows. As discussed in section 5.2, the variability of stream macrozoobenthos has repeatedly been found to be significant at the reach scale, which makes within-reach variability highly relevant for the present analysis.

At the same time, management under the WFD requires stream water bodies to be able to host local populations of these organisms. In Germany, the spatial extent of a WFD water body is on average 14.4 km (www.umweltbundesamt.de, consulted on June 10, 2014). The reach-scale referenced here therefore refers to sub-segments of a WFD water body, varying in length between a few hundred meters and a few km depending on the size of the stream.

This is the spatial extent of the habitat simulation strategy proposed in this thesis. This choice of spatial extent allows capturing enough space that habitat for several local groups of individuals may be modeled.

The justification for this is that the survival of the target species in the reach is unlikely to be linked to any particular group of individuals inhabiting a small suitable area. Rather, the method aims at assessing hydromorphological quality by identifying habitable space areas over an entire reach, which is much more suitable for organisms with a strongly patchy distribution such as macrozoobenthos.

A similar logic is followed in setting the **temporal extent** of the analysis. Given that the goal under the WFD is the long-term persistence of type-specific taxa in the reach, the habitat simulations proposed in this thesis cannot be restricted to any specific discharge conditions. Rather, they must encompass several generations of the target species, which also follows from the principles discussed in sections 5.4 and 5.3 of the conceptual basis.

Therefore, several hydrological years must be simulated in order to apply the strategy formulated here. This temporal extent exceeds typical simulation times in river engineering. However, alternative methods in fluvial geomorphology such as 'reduced complexity models' allow overcoming this technical obstacle (Bokulich 2013, Coulthard et al. 2007, Coulthard and Van De Wiel 2012) while maintaining sufficient accuracy. This issue is further discussed in sections 6.2.2 and 8.2.4.

The **spatial resolution** of the habitat simulations aims to imitate the biological sampling scales normally applied in stream macrozoobenthos studies. In this field, macroinvertebrate samples taken on (typically) 25 cm by 25 cm quadrats are related to measurements of abiotic variables (velocity, depth, substrate) done at the same or similar scales.

Table 6.1: Criteria concerning the selection of spatial and temporal scales for describing the hydromorphological template.

	Resolution	**Extent**
Space	• Aggregation of small-scale dynamics • Scale of biological sampling • Relation between micro-scale (mm-cm) and meso-scale (tens of cm) flow	• Population-level persistence • Groups of individuals instead of individual organisms
Time	• Hydrological dynamics of stream • 20%-50% smaller than scale of fastest changes (O'Neill et al. 1996)	• Persistence of target species over several generations

Following this method, stream ecologists have been able to find significant habitat associations for many species of macrozoobenthos, which suggests that there is indeed a relationship between these organisms and habitat at this scale. Therefore, the description of the hydromorphological template proposed in this thesis is based on hydrodynamic models and substrate information at this level of resolution.

However, the argument could be made that the scales of perception of small animals such as macrozoobenthos, which range in size between a few millimeters and a few centimeters, are also of this order, and that spatial aggregation over tens or hundreds of squared centimeters eliminates the abiotic gradients actually experienced by them.

Although it is true that such spatial aggregation leads to loss of detail, any science studying complex systems requires aggregation of some kind in order to connect small-scale mechanisms with management-scale effects (see section 5.2).

In ecology, this is achieved by working with aggregate variables that are 'reasonable extensions of known or suspected processes at lower levels' (Orians 1980), which is what the sampling quadrats mentioned above are expected to represent. In turn, modeling habitable space with a spatial resolution of a few tens of centimeters and up to a meter (depending on the size of the stream) aims to aggregate the micro-scale flow conditions experienced by individual organisms.

As long as there is a correspondence between flow at the scale of a few tens of centimeters and micro-scale flow (mm - cm), the error introduced by this aggregation will remain small compared to other uncertainties. In fact, the relationship between spatially-averaged and small-scale (near-bed) flow conditions is well-established in open channel hydraulics, and methods exist for linking the two in order to obtain a closer description of the micro-scale flow near the bed (Bezzola 2002) (see section 7.2.2 for further discussion).

Finally, the **temporal resolution** of the habitable space simulation is given by the hydrological dynamics of the stream. This is a central aspect of the method, since it is these temporal variations that lead to the variability that makes up the hydromorphological template.

Following principles from landscape ecology (O'Neill et al. 1996), the temporal resolution should be between 20% and 50% smaller than the time scale of the fastest changes.

Table 6.1 summarizes the above considerations regarding extent and resolution of analysis for the hydromorphological template.

The next section addresses the implications of this selection of spatial and temporal scales for the hydrodynamic models underlying the habitat simulations.

6.2.2 Are high-resolution 3D models necessary?

As mentioned above, the habitable space analysis proposed here is based on 2D shallow water numerical simulations of the studied reach. However, it could be argued that only a high-resolution three-dimensional model would be able to depict the complex flow patterns that exist in the boundary layers inhabited by stream macroinvertebrates (Nikora 2010).

Nonetheless, further consideration indicates that ecological knowledge on these organisms' habitat associations is based on descriptions of the hydraulic environment in broad terms, e.g., 'slow-' or 'fast-flowing' areas, 'flow velocity below 0.4 m/s', etc. (see for instance http://www.freshwater-ecology.info/, or the expert rule sets compiled by Kopecki and Schneider 2010).

Thus, in order to examine habitable space dynamics with this ecological information, what is required is a hydraulic model that describes the hydrodynamic characteristics of the study area in these terms, i.e., that allows identifying relevant flow patterns on scales between a few tens of centimeters and a few meters, depending on the size of the stream (see section 5.2).

Any analysis done at a higher resolution will concentrate on individual activity on shorter time spans. In eco-hydraulics, this involves processes such as the behavioral response to drag, turbulence and shear and its relation to different substrate types (Rice et al. 2008).

Biological modeling at this scale is commonly done in the form of individual-based models, which aim to predict individual movement and behavior based on environmental conditions (flow velocity, turbulence, temperature, oxygen) and their associated physiological stress (Li et al. 2010, Rochette et al. 2012).

However, arriving at management-relevant conclusions requires population-, community- or ecosystem-level inferences, which means individual-based results must be *upscaled* in space and time. In ecology, though, upscaling requires deep knowledge of the interactions between environmental factors at different scales (Schneider 1994), and much of this knowledge is still lacking (Downes 2010).

In addition, the small-scale (e.g., millimeter-level) flow field in each of these flow areas is physically tied to the prevailing velocity and substrate at the intermediate scale mentioned above (a few tens of centimeters, a few meters).

This connectivity between scales is given through hydrodynamics: it is possible to obtain a good picture of micro-scale flow conditions, which are the ones actually experienced by macroinvertebrates, from a meter-scale-resolution shallow water simulation that allows identifying contrasting flow areas such as pools, riffles, velocity and depth gradients at a bend, etc.

Moreover, these inter-scale linkages are commonly used in hydrodynamics, for instance to estimate bottom shear stress for flows over rough or vegetated beds (Baptist 2005, Bezzola 2002). Thus, flow velocities calculated with a depth-averaged model may be adjusted using such approaches, leading to a more accurate description of habitable space (Kopecki 2008).

Therefore, for the purpose of describing the hydromorphological template, a millimeter-resolution three-dimensional model may provide the same amount of ecologically-relevant information as a shallow-water model of, say, 0.5- or 1-m resolution with an adjustment method for bottom velocity in complex substrate areas.

For a detailed description of the application of this adjustment technique, see section 7.2.2.

6.2.3 Habitable space patches

As discussed in section 6.1, the overlay of substrate, flow velocity, water depth and tolerance ranges results in a habitable space map for each time step in the analysis. Each of these maps is a raster in which a suitability value indicates whether a pixel is habitable or not.

In the method proposed here, habitat suitability is treated as a binary ('habitable', 'inhabitable') rather than as a continuous (e.g., an index between zero and one) or categorical (e.g., 'low', 'medium' and 'high') variable. This follows from the discussion presented in section 5.1, where it was noted that responses at the population-level are by nature threshold-type responses.

On this basis, it is more appropriate to represent tolerance ranges being exceeded by using a binary scheme that classifies abiotic conditions beyond tolerance as inhabitable, and vice versa.

Each habitable space map will therefore consist of habitable and inhabitable pixels, which may be clumped together into **patches** of habitable space based on whether they are adjacent or not.

Thus, in this dissertation, a patch is simply a continuous surface of habitable space surrounded by inhabitable area.

This patch definition follows from the patch dynamics concept of streams (Pringle et al. 1988) (see section 5.4), where 'a patch is a spatial unit that is determined by the organism(s) and problem(s) in question'. It is also important to note that 'patches are not fixed elements of the landscape, but rather are useful spatial constructs that vary with the objectives of a given study' (Turner et al. 2001).

In this dissertation, the patch perspective is applied to habitable space with the objective of describing the studied stream in a way that can be related to the long-term survival chances the target species has in it.

For this, it is necessary to identify all areas within the stream that can be inhabited at any given time, and then track their behavior throughout several generations of the target species (see definition of hydromorphological template in section 6.1).

Building patches out of individual pixels is a natural step in this process, and is a technique commonly applied in the field of terrestrial population ecology for studying population survivorship at the landscape scale (Akçakaya 2000).

Each of the maps that results from this method contains a collection of habitable patches in the study area, which is sometimes referred to as the 'patch network' (Hanski 1999a) of the area, or, in this case, the stream reach.

Section 7.6 presents the technical details of the patch-building process discussed here.

6.2.4 Analysis variables: patch dynamics indicators

Until now, the proposed approach has provided a set of habitable space maps, one for each time step, in which adjacent habitable pixels are clumped together into habitable space patches. The next step in the process of describing the hydromorphological template is therefore summarizing this information by implementing the principles of the patch dynamics perspective (Poole 2002, Pringle et al. 1988, Townsend 1989).

For this purpose, the analytic framework proposed by Pringle et al. (1988) is adopted, which describes patch dynamics in terms of **patch size and size distribution**, **patch density**, **patch juxtaposition** and **patch duration**. Consequently, the indicators presented in this section address each of these aspects of patch dynamics.

The indicators are based on the principle that variability in habitable space, which results in temporal or permanent reductions in the size of the identified patches, is a phenomenon that increases the chance of individuals being removed from the target population, either by death or by drift.

Thus, it is important to characterize the intensity with which such potential removal events take place in order to gain insight into the potential hydromorphological stress to which the target population is exposed.

The discussion in this section concentrates on the nature and purpose of each of the indicators, which are presented here in order to indicate the possibilities offered by the use of numerical shallow water models for the ecological analysis of hydromorphology.

In order to judge these indicators, bounds must be provided for their values such that the reach can be characterized as hydromorphologically suitable or unsuitable. The rationale for this is presented in the third element of the method (section 6.3). The proof of concept presented in chapter 8 is an empirical test of this rationale.

The technical details of the computation of the indicators are discussed in chapter 7.

6.2.4.1 Patch area and area-duration curves

The basis of the following analyses is provided by the patch building and patch tracking processes described in sections 7.5 and 7.6. The outcome of these processes is a matrix containing the areas of each of the patches that existed in the stream reach during the simulation period ($A_{i,t}$ in equation 6.1), which can then be used to compute the indicators proposed in this section.

The total habitable area available to the target species in the study reach at time t, A_t, is simply the sum of the areas of the individual patches $A_{i,t}$:

$$A_t = \sum_{i=1}^{N_t} A_{i,t} \tag{6.1}$$

with N_t = number of patches at time step t.

Careful consideration must be given to the resolution of the spatial data (substrate maps, hydrodynamic numerical models) used in this analysis in order to reduce errors in the estimation of patch areas (Crowder and Diplas 2000) (see sections 5.2 and 7).

The abundance (number) of habitable patches at time t, N_t, can be readily derived once patch areas have been computed by simply allocating a 1 to all A_i values greater than zero and summing across all patches. This will produce a time series of N_t of length equal to the number of time steps in the analysis.

Habitable space variability can be assessed for each individual patch using the areas resulting from the previous steps of the method (for technical details, see chapter 7). The coefficient of variation of patch area can be defined as

$$cv(A_i) = \frac{\sqrt{\frac{\sum_{t=1}^{T}(A_{i,t}-\overline{A_i})^2}{T}}}{\overline{A_i}} \tag{6.2}$$

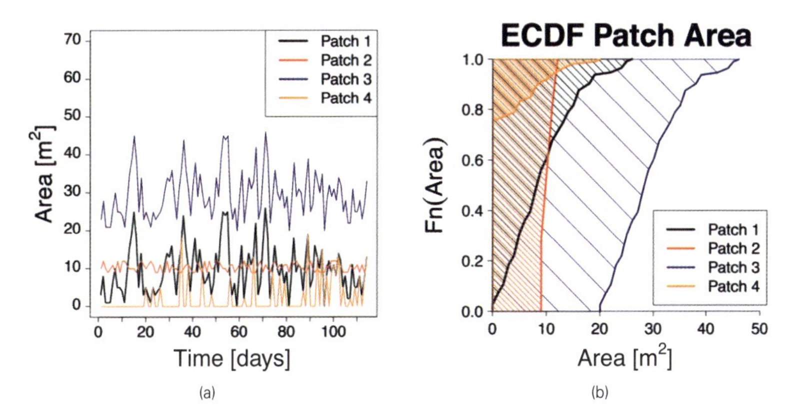

Figure 6.5: (a) Area time series for four hypothetical patches. (b) Empirical cumulative distribution functions (ECDF) of the four hypothetical patches in (a). Fn (y-axis in b) indicates the cumulative relative frequency of the observations. Shading in the corresponding color indicates the area above each ECDF curve in (b).

where cv = coefficient of variation, $A_{i,t}$ = area of patch i at time t, T = total number of time steps and $\overline{A_i}$ = average area of patch i. The numerator of this fraction corresponds to the standard deviation of patch area.

Variability in patch sizes and duration can also be described by constructing area-duration curves for each patch. These curves are based on the empirical cumulative distribution function (ECDF) of the computed patch areas.

For any given patch, the time series of patch area is sorted in ascending order and the fraction (0 - 1) of observations at or below each point is determined. Thus, area-duration pairs are constructed, with duration being represented by the fraction of the simulation time in which areas equal or smaller than the corresponding area were observed.

This process results in an ascending curve that rises towards one at different rates depending on the patch's behavior. For a patch featuring mostly small sizes throughout the simulated period, the ECDF curve will rise rapidly and the area above it will be correspondingly small. In contrast, a large patch with only minor oscillations will exhibit an ECDF curve that rises slowly towards one, thus having a larger area above it.

Thus, the area above a patch's area-duration curve gives a measure of the availability of habitable space provided by the patch, simultaneously in terms of patch size and duration. This quantity is proposed here as a measure of the availability of habitable space afforded by the patches existing in the stream. At the same time, it provides a means for comparing the suitability of the simulated habitable space patches.

Figure 6.5a presents the area time series of four hypothetical patches. Patch 1 has an average area of 9.6 m^2 and exhibits large oscillations, occasionally reaching sizes of more than 20 m^2 but dropping to near-zero values immediately afterwards.

Patch 3 exhibits the same temporal trend as patch 1, but with an average size of 29.6 m^2 and a minimum area of 20 m^2. In contrast, patch 2 has an average size of 10 m^2 and much smaller size oscillations (range = 3 m^2). Patch 4 represents a very harsh environment, with a mean area of 2.1 m^2 and very frequent zero values interspersed with occasional increments of up to 20 m^2.

The ECDF curves for these four patches are presented in figure 6.5b. Patch 4 (orange) features the ECDF curve with the smallest overlying area (2 m^2), which is consistent with its small and occasional supply of habitable space. In contrast, patch 3 (blue) never drops in size below 20 m^2 and, despite its size variations, offers a permanent supply of habitable space (area above = 29.1 m^2).

This is similar to the situation in patch 2 (area above = 10 m^2), although its minimum size is 10 m^2. Patch 1 (area above = 9.1 m^2) represents an environment that is intermediate between patches 4 and 2, with an area greater than 5 m^2 more than 50% of the time but several drops to near-zero values throughout the analysis period.

The fact that the area above the curve for patch 1 is close to that for patch 2 is problematic, since these two patches represent very different scenarios. This highlights the need to account for the duration of patch size oscillations explicitly, which is approached in section 6.2.4.4.

A similar analysis would be possible using absolute instead of relative frequencies, which would lead to an indicator with dimension $T \cdot L^2$ instead of L^2.

6.2.4.2 Reach-scale habitable space losses - I_Δ

The spatially and temporally explicit information generated by the proposed approach also allows a detailed examination of potential stress (*sensu* Underwood 1989) due to losses in patch areas. Contractions in the size of individual patches can be understood as generating potential stress to the target species, since individuals may be removed from local populations through death and/or drift.

In this section, an indicator that summarizes this potential stress at the reach level is proposed in order to illustrate the possibilities that arise when the existing ecological knowledge is combined with numerical shallow water models. The development of this indicator, denoted here as I_Δ (see below), considers 1) the size of individual patch area losses throughout the reach and 2) the size of each patch at the time of the area contraction.

The discussion in this section focuses on the rationale for such an indicator and its characteristics. Unfortunately, a direct empirical check of its performance was not possible with the available data. However, the proof of concept presented in chapter 8 can be understood as a successful test of its underlying logic.

Specifically, the fact that sensitive taxa do exhibit limitation at low values of habitable space availability (see chapter 8) suggests that an indicator that summarizes habitable area losses would show a similar behavior. That is, a reach with a low habitable space availability would be similar to a reach comprised of patches experiencing large and frequent area losses. Nonetheless, it is acknowledged that a direct test of I_Δ is the only way of proving this conclusively.

The rationale for I_Δ is as follows. If we look at an entire reach, large relative size losses in large patches may still leave, in probabilistic terms, enough space for local populations to persist. On the contrary, a reach in which habitable space is distributed in small patches experiencing large area losses is potentially more stressful, since very little habitable space will be left after each loss event.

On this basis, I_Δ is computed as

$$I_\Delta = \frac{\sum_{i=1}^{N_t} \frac{1}{A_{i,t}} \frac{(A_{i,t+\Delta} - A_{i,t})}{A_{i,t}} V}{\sum_{i=1}^{N_t} \frac{1}{A_{i,t}}} \tag{6.3}$$

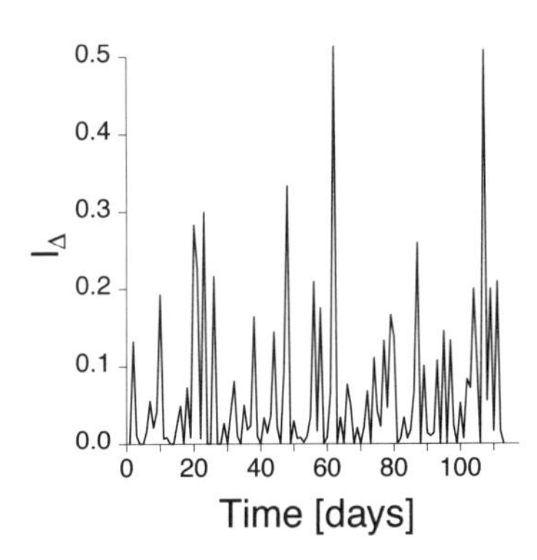

Figure 6.6: I_Δ time series for the hypothetical network made up of the four patches of figure 6.5.

where i denotes patch i of the network and Δ is the time step for the analysis. The selection of time interval Δ must follow biological criteria related to the tolerance of the target species to habitable space losses, as given by its behavioral, physiological and life-history characteristics (e.g., Schmidt-Kloiber and Hering 2012).

v is a factor that eliminates positive area changes (i.e., area gains, see below), and is defined as

$$v = \begin{cases} 0 & A_{i,t+\Delta} - A_{i,t} \geq 0 \\ -1 & A_{i,t+\Delta} - A_{i,t} < 0 \end{cases} \tag{6.4}$$

In equation 6.3, factor $1/A_{i,t}$ accounts for the effect of patch size on the probability of local extinction, which becomes smaller as patch area grows. In turn, $(A_{i,t+\Delta} - A_{i,t})/A_{i,t}$ accounts for the effect of the relative size of the area loss experienced by the patch.

Using the hypothetical 4 patches from the last section, the above equations lead to the time series of I_Δ plotted in figure 6.6. In this hypothetical network, I_Δ exhibits a very erratic distribution, with several small, medium and large peaks throughout the study period.

Multiplying each relative area loss by the inverse of the absolute size of the patch ($1/A_{i,t}$) allows accounting for the fact that the chances of local extinction are inversely related to the size of the patch (Hanski 1999b), and therefore larger remaining areas may host local groups of individuals less prone to extinction.

Note that whenever the area of a patch is zero, it is simply left out of the analysis given that a zero-area patch cannot, by definition, experience area losses. This excludes the possibility of division by zero in the calculation of I_Δ.

However, a patch with zero area also represents a source of stress, since it does not offer any habitable space at that time step and may do so only occasionally (e.g., patch 4 in the last section). This dimension of the hydromorphological template is captured by the analysis of patch duration (section 6.2.4.4).

It is also important to mention that the factor $1/A_{i,t}$ drops very rapidly with patch area. This is intended to represent the fact that potential stress due to small patch size lets up very quickly with increasing patch size for organisms as small as macrozoobenthos, whose perception scales are much smaller than the patches identified in the analyses proposed here (Wiens 1989).

Gains in patch area between t and $t + \Delta$ are excluded because potential stress due to habitat losses is not ameliorated by them. It is highly unlikely that macrozoobenthos are able to locate and exploit newly available habitable space over a time period of a few days, which is the scale of Δ values intended in this analysis.

Furthermore, most of the macrozoobenthos species used in assessment under the WFD are insects, whose aquatic larvae do not reproduce in the stream. This excludes the possibility that increases in aquatic habitable space lead to increases in population density and therefore also to increases in the species' long-term survival chances.

6.2.4.3 Point-based reach-scale habitable space availability - *hsAv*

The aforementioned analyses of habitable space are based on continuous substrate information such as substrate maps (e.g., raster), which show the location and extent of the different substrate types in the studied stream reach.

However, in many cases, such data are only available as punctual observations. This type of substrate data consist of a series of points distributed throughout the reach and a classification of the predominant substrate type at each point (Schröder et al. 2013), which is typically done in the context of macrozoobenthos sampling campaigns.

With these data, the estimation of aquatic habitable space must be adjusted to allow for the overlay of spatially continuous velocity and water depth data (from the hydrodynamic model) with the punctual substrate observations (figure 6.7). The main reason for this adjustment is that interpolating the punctual substrate data into a continuous surface (e.g., raster) is very likely to lead to inaccurate estimations of the actual substrate areas, since the actual shape of substrate patches is unknown.

Therefore, an alternative for assessing habitable space based on punctual data is proposed. With this alternative, it is no longer possible to estimate patch areas, and, correspondingly, to compute the indicators described in sections 6.2.4.1 and 6.2.4.2. However, this strategy still allows assessing the reach-scale availability and variability of habitable space, and is more reliable than habitable area estimations based on uncertain interpolated substrate data.

Furthermore, given that punctual substrate information is typically gathered with the aim of having a representative sample of the reach, which is necessary in macrozoobenthos sampling (Hering et al. 2003), it is possible to consider these punctual data as a representative sample of the habitable space availability of the reach.

The result of the overlay of continuous and punctual information is a set of points, each of which can be either habitable or inhabitable (figure 6.7) at each time step of the analysis depending on the habitat associations of the target species.

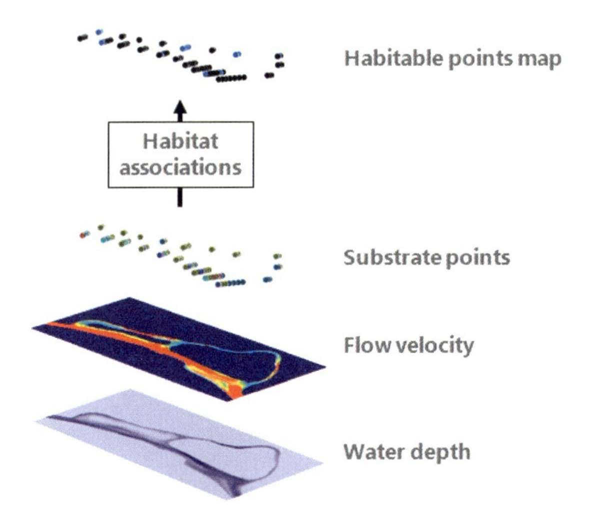

Figure 6.7: Conceptual diagram showing the overlay of continuous water depth and flow velocity data with punctual substrate observations. In the water depth map, darker blue areas indicate deeper water. In the velocity map, red areas indicate higher flow velocity, whereas blue areas indicate lower values. In the substrate points map, each color denotes a different substrate type. In the habitable points map, blue = habitable, black = inhabitable. Data from site wr (see section 8.2.1 for a detailed description).

With these habitable points maps it is possible to obtain a time series of the estimated fraction of habitable space in the reach (figure 6.8a), which results simply from dividing the number of habitable points by the total number of observation points at each time step.

This time series can then be analyzed in a similar fashion as patch areas in section 6.2.4.1. Namely, an ECDF (empirical cumulative distribution function) curve can be constructed which summarizes the spatiotemporal availability of habitable space in the reach. As for patch areas, the area above the curve can be used as an aggregate measure of habitable space availability in the stream (figure 6.8b).

This quantity is denoted here as **hsAv**. Figure 6.8 shows an example in which this indicator is used for distinguishing between the habitable space supply for the species *Odontocerum albicorne* (Trichoptera) in a trained (engineered, black) and a restored (grey) reach on the river Lahn (Federal State of Hessen, Germany) over a period of 5 years (2005 - 2009).

The restored reach exhibited a supply of habitable space three orders of magnitude greater than the trained reach, as shown by the computed hsAv values. These data are part of the information used in the proof of concept of the approach. For a complete description, see chapter 8.

6.2.4.4 Patch duration

The total duration of patch i, d_i, quantifies the number of time steps in which that habitable patch existed throughout the simulation period. d_i is defined as:

$$d_i = \sum_{t=1}^{T} 1 \cdot h_{i,t} \tag{6.5}$$

where T = number of time steps of the simulation period and $h_{i,t}$ is a factor that accounts for whether patch i is present at time step t. $h_{i,t}$ is defined as

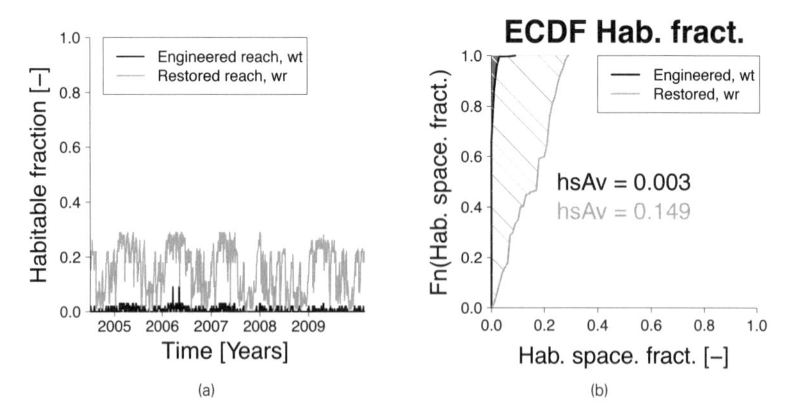

Figure 6.8: (a) Time series of habitable space fraction for the macrozoobenthos species *Odontocerum albicorne* (Trichoptera) in an engineered (wt) and a restored (wr) site on the river Lahn (Federal State of Hessen, Germany) near the city of Wallau. (b) ECDF (empirical cumulative distribution function) curves of habitable space for the same two sites. For further details on the data, see section 8.2.1.

$$h_{i,t} = \begin{cases} 1 & A_{i,t} > 0 \\ 0 & A_{i,t} = 0 \end{cases} \quad (6.6)$$

The existence of a patch can be fragmented into several disconnected **existence spells**, which are separated by time periods in which the patch disappears entirely from the reach due to inadequate water depth, flow velocity or substrate. Therefore, statistics that describe the number and length of existence and disappearance periods are necessary in order to characterize patch duration.

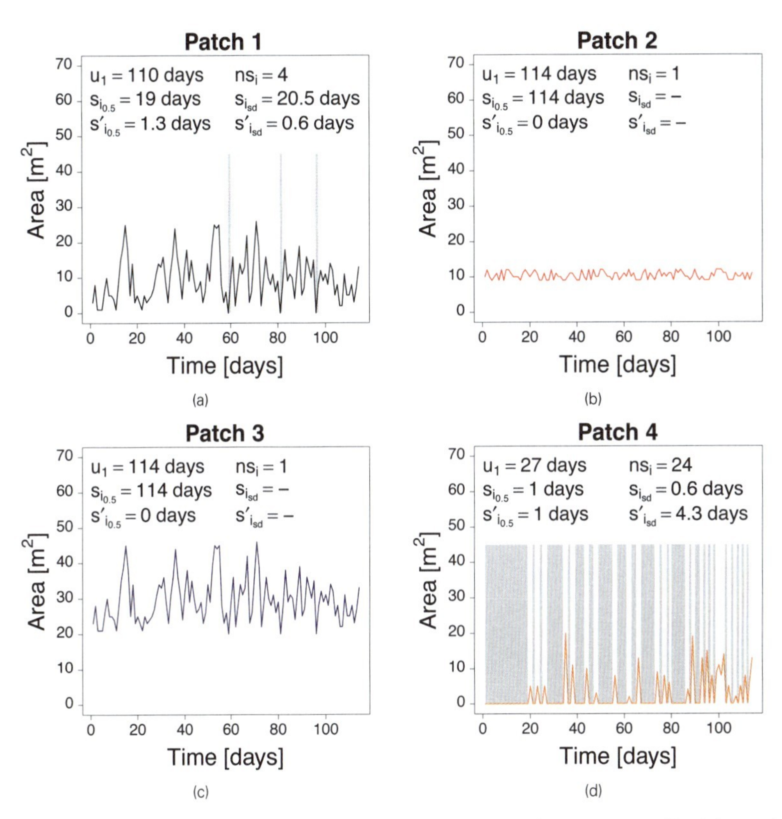

Figure 6.9: Patch duration indicators for the four hypothetical patches of section 6.2.4.1. Shaded areas in each plot indicate time periods in which patch did not exist. Note that $s'_{i0.5}$ and s'_{isd} do not apply to patches 2 and 3 since they never disappear. For symbols, see table 6.2.

Table 6.2: Patch duration indicators

Symbol	**Indicator**
d_i	Total duration of patch i
ns_i	Number of existence spells for patch i
$s_{i0.5}$, s_{isd}	Median and standard deviation of existence spell duration for patch i
$s'_{i0.5}$, s'_{isd}	Median and standard deviation of disappearance period duration for patch i

As for any continuous quantity, several descriptive statistics are available for characterizing patch duration. The ones presented here are simply a subset that aims to capture its most relevant aspects according to the principles outlined in the conceptual basis. The proposed statistics are summarized in table 6.2.

The values of these indicators for the four hypothetical patches presented in section 6.2.4.1 are summarized in figure 6.9.

6.2.4.5 Patch juxtaposition

Finally, the analysis of patch juxtaposition proposed here aims to describe the overall separation between habitable space patches, which can provide insight into how adequate the spatial configuration of habitable space is with respect to processes that involve movement of individuals, such as within- and inter-patch journeys (Bond et al. 2000, Lancaster 2008, Lancaster and Downes 2013).

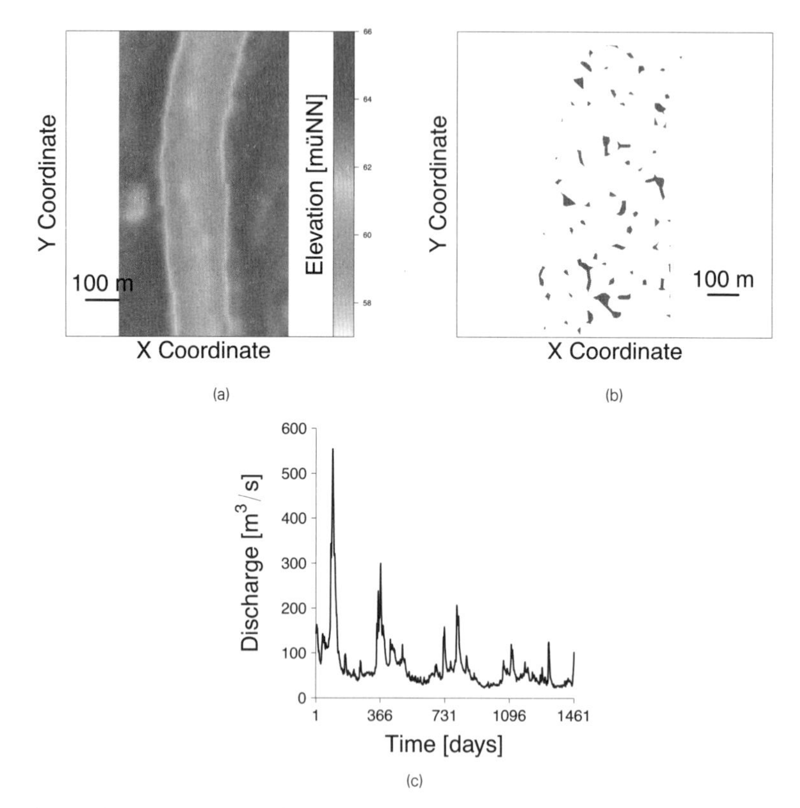

Figure 6.10: Hypothetical (a) Digital Elevation Model, (b) suitable substrate map and (c) discharge time series used for the exemplary computation of the potential connectivity indicator *c*. See text for further description of the data.

Although a large number of indicators are available in areas such as metapopulation (e.g., Hanski 1999b) and landscape (Turner et al. 2001) ecology, the approach proposed here follows a simplified strategy that aims to exploit the detailed flow field information provided by the numerical shallow water model.

On this basis, a connectivity indicator is proposed based on the proportion of random, hypothetical, passive (hydrochoric) dispersal events that finish within habitable space. This indicator is denoted here as c.

The algorithm (see chapter 11) computes the probability that a random hypothetical dispersal journey that starts within habitable space also ends at a suitable point. This probability is calculated as the fraction of all hypothetical dispersal journeys that finish within habitable space.

First, ten random points are placed within each habitable space patch in the study reach. Then, for each of these starting points at each patch, a journey length is drawn randomly from the dispersal kernel of the target species, which is a negative exponential function (Bond et al. 2000) relating dispersal distance and its probability density.

The key parameter of this exponential function is its decay rate α, which summarizes what is known about the dispersal capacity of the target species. The path followed during the dispersal events is calculated using the flow direction from the 2D hydraulic model.

If a journey ends within a patch of habitable space, it is counted as a successful dispersal event. Similarly, if a journey touches a habitable space patch other than the source patch (i.e., the patch where the journey started), it is also counted as a successful dispersal event.

The result is an indicator of **potential** connectivity (*sensu* Calabrese and Fagan 2004) for each time step, and it is interpreted as the probability that a random hypothetical passive journey starting within a habitable space patch also ends within habitable space.

In order to implement the algorithm for the calculation of c (see chapter 11), a hypothetical river reach was simulated based on data extracted from an existing numerical shallow water for the river Elbe (Bleyel and Faulhaber 2007). This model was obtained in the context of an inter-institutional agreement between the TU Dresden's Institute of Hydraulic Engineering and the German Federal Institute of Hydraulic Engineering (Bundesanstalt für Wasserbau, BAW).

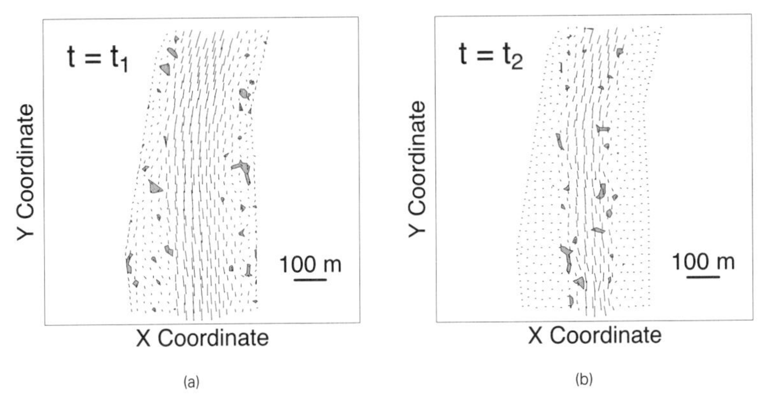

Figure 6.11: Habitable patch network and overlying velocity field resulting from the LISFLOOF-ACC simulation of the hypothetical river reach used in the computation of c. Two different time steps are shown (t_1 and t_2). The length of the arrows is proportional to flow velocity, which was calculated using hypothetical data.

A small segment of the Elbe model was extracted and re-interpolated in order to generate a Digital Elevation Model that included the river channel and a small adjacent floodplain area (figure 6.10a). Hypothetical substrates were created using R's capabilities for random number generation, and a subset of these substrates were then designated as suitable in order to generate a suitable substrate map (figure 6.10b).

An artificial discharge hydrograph (figure 6.10c) was generated that imitated discharge dynamics at the Elbe gauging station Wittenberg of the German Waterway and Navigation Administration (Wasserstraßen- und Schifffahrtsverwaltung des Bundes, WSV). The discharge time series at this gauge was simply multiplied by 0.3 in order to adjust the water quantity to the reduced hypothetical model. This was done with the purpose of obtaining an artificial time series that was adequate for the hypothetical reach but still retained a real discharge dynamics, so as to provide some degree of realism to the hypothetical example.

These data were used as input for the shallow water model LISFLOOD-ACC (a detailed description of LISFLOOD-ACC is presented in section 8.2.4), which was used to simulate 4 years of discharge in the hypothetical reach. The results of this computation were overlain with the hypothetical substrate data in order to produce an dynamic habitable patch simulation. The resulting set of patch network maps (see examples in figure 6.11) were then used for computing *c*.

The resulting time series for *c* is shown in figure 6.12. Although computed with hypothetical data, the fact that a realistic discharge dynamics was used allows drawing some general conclusions about *c* in real systems.

High discharge peaks seem to drive potential connectivity down, which is likely the result of a lower availability of habitable space during such periods, when fewer dispersal journeys can be expected to end within habitable space. At the same time, higher *c* values during lower discharge conditions suggest that, whenever discharge is low but still above a certain threshold, potential connectivity remains high.

It is important to note that the goal of this analysis is not to predict the actual trajectory followed by dispersing individuals of the target species. Rather, it is an attempt to provide a simple measure of potential connectivity that is suited to the organisms and scale of analysis at hand, that can be interpreted and integrates well into the proposed methodological framework, and that captures existing ecological knowledge.

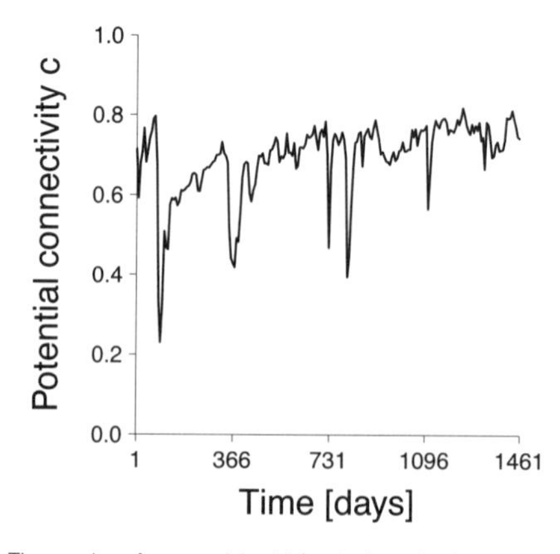

Figure 6.12: Time series of connectivity (*c*) for the hypothetical reach of figure 6.10.

Clearly, dispersing macroinvertebrates do not always behave as passive particles during dispersal (Lancaster 2008), and dispersal kernels are very likely to be site-specific depending on the distribution of obstacles in the flow (for instance, dead water zones, Bond et al. 2000).

6.3 ELEMENT 3: MEASURING THE LIMITING ROLE OF HYDROMORPHOLOGY

Up to this point, several indicators have been proposed by combining the principles from the conceptual basis and the spatially and temporally explicit information provided by unsteady habitable space simulations. These indicators summarize the availability of aquatic habitable space and provide a detailed picture of the hydromorphological conditions to which the target biological populations are exposed in the stream.

The next step in the development of the method is to address the question of how little habitable space can be considered limiting. However, this question is strongly site- and species-specific. In the same sense that a Minimum Viable Population for a species is hardly transferable between environments and local populations (Boyce 1992), a universally-valid limitation threshold for a species would be very difficult to derive.

In attendance to this, the proof-of-concept presented in chapter 8 concentrates on proving the existence of this limiting relationship with empirical data, understanding that the limitation threshold that results from this process is restricted to the studied river section, time period and macrozoobenthos community.

Nonetheless, confirming the existence of this limiting-type relationship between the proposed indicators and macrozoobenthos allows verifying the proposed habitat modeling approach and proves it has a sound ecological basis. At the same time, this proof-of-concept can be seen as a confirmation of the ecological and fluvial hydraulics principles on which the method is based.

Although these ecological concepts represent syntheses of large bodies of empirical evidence, any implementation of them requires a case-specific confirmation, given that they are, unlike principles in areas such as physics, abstractions made at a very high level of complexity (Pickett et al. 2007).

As described in chapter 8, the proof-of-concept allowed proving the hypothesis that a limiting effect would be detected for macrozoobenthos taxa expected to be more sensitive to habitable space scarcity, whereas such limitation would be much weaker for more tolerant taxa. The fact that this limitation could be confirmed proves that the proposed indicators do provide a description of stream hydromorphology that is relevant for the long-term survival of the target species, which is the goal of the approach.

Moreover, these results are in agreement with the central tenet that unifies the physical habitat template (Poff and Ward 1990), habitat template theory (Southwood 1977) and patch dynamics in streams (Pringle et al. 1988, Townsend 1989), namely that the time course of habitat availability in space and time relative to specific life cycle requirements is a critical factor for the long-term persistence of biological populations, particularly in dynamic environments (Levins 1962, Thiery 1982).

A detailed description of the approach followed in the proof-of-concept, and of the data and methods used, is presented in chapter 8.

7 TECHNICAL ASPECTS

This chapter presents an overview of the most important technical aspects of the approach. It aims to show how the indicators proposed in the previous chapter can be computed based on numerical shallow water simulations, substrate maps and tolerance information about the target species.

7.1 GENERAL DESCRIPTION OF THE APPROACH

The method proposed in this thesis uses unsteady shallow water hydrodynamic simulations and substrate mapping tools to produce ecologically meaningful information about the hydromorphology of a stream reach. The analysis can be broken down into the following steps (figure 7.1)

1. Selection of target species
2. Substrate mapping and two-dimensional hydrodynamic simulation
3. Formatting of input data as described in chapter 11
    - Water depth and velocity (from hydraulic model)
    - Substrate map
4. Habitat suitability assignment
5. Patch building
6. Patch tracking
7. Calculation of patch dynamics indicators

7.2 HYDRAULIC SIMULATION

The key technical element of the proposed approach is an unsteady shallow water numerical hydrodynamic model. This model allows generating detailed, spatially and temporally explicit habitat information, which constitutes the basis for the description of the hydromorphological template.

Shallow water models are based on the depth-averaged form of the three-dimensional, Reynolds-averaged, incompressible Navier-Stokes equations (Toro 2001). This type of hydrodynamic numerical model is able to capture spatial and temporal flow patterns at the scales required for describing the hydromorphologicl template (see section 6.2).

The depth-averaged flow velocities used in the proposed method do not aim to represent exactly the small-scale flow conditions in the interstices inhabited by macrozoobenthos. Rather, the shallow water simulation is intended to describe stream habitat at the scale at which significant associations have been found with the macrozoobenthos community, i.e., the scale of the sampling devices used in stream ecology. A detailed discussion on this is presented in section 6.2.1.

However, in order to account for the fact that natural riverbeds are typically rough due to the presence of dead wood, boulders, roots and other structures, the adjustment method of Bezzola for rough riverbeds is employed (see section 7.2.2).

7.2.1 Model extent and resolution

The first component of the approach presented here are the results of a long-term simulation of the study reach's hydrodynamics. Hydraulic simulation tools were first used in running water ecology in the early stages of physical habitat modeling (e.g., Bovee 1982), and have become an integral part of many recent assessment methodologies (e.g., CaSiMiR-Benthos [Kopecki and Schneider 2010], INFORM-MOBER [Giebel et al. 2011]).

Here, however, hydrodynamic simulations are used with a different focus in order to provide input for analyses of habitat dynamics in the context of the physical habitat template (Poff and Ward 1990). According to this ecological concept, habitat dynamics, in terms of spatiotemporal variability of substrate and flow, is a more relevant environmental force for stream populations than punctual values of these variables in space and time.

Hence, the spatiotemporal resolution and extent at which this variability is described is a key aspect of the approach, which has direct implications for the resolution and extent of the hydraulic models used in the analysis. As discussed in section 6.2.1, the required extent and resolution are given by the ecological characteristics of the target species and the resources it uses, i.e., the patches of aquatic habitable space.

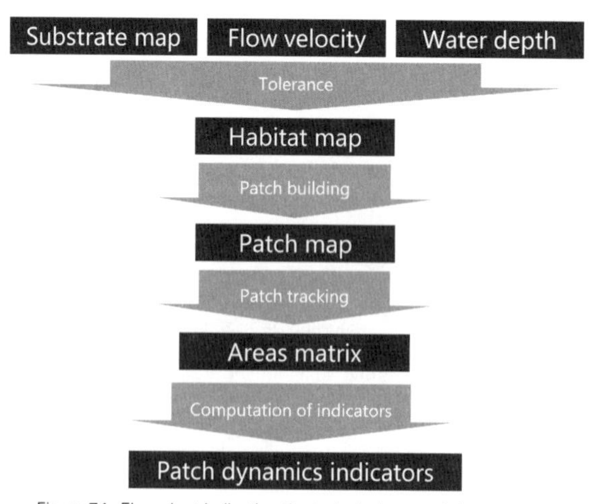

Figure 7.1: Flow chart indicating the technical steps of the approach.

7.2.1.1 Proposed resolution and extent in this approach

For an ecological phenomenon such as the inter-annual persistence of macroinvertebrate populations in a stream reach, the above discussion implies that flow conditions *at or below the meter-scale* are necessary over a reach that encompasses a collection of habitat patches, as a rule of thumb, from 1 to 10 times the flow width.

Using this level of spatial resolution is necessary given that averaging flow quantities over an entire cross section (as in 1D models) or over 2D model elements tens of meters in size (a common mesh resolution in many hydraulic engineering applications) may obscure relevant variations in velocity and substrate. The use of 1D models is therefore not recommended in this approach, a view shared in recent discussions on eco-hydraulics (e.g., Lane and Ferguson 2005).

2D shallow water models may be better suited for this task (Crowder and Diplas 2000, Lane and Ferguson 2005), although model resolution must be carefully managed in order for the resulting velocity field to capture ecologically-relevant flow features (Crowder and Diplas 2000) (see section 6.2.1). Among these are flow patterns around obstacles, which are both common and necessary in river systems of a high ecological condition (e.g., large woody debris, boulders, cobbles).

The temporal resolution is also given by the rate at which the ecological process takes place. In this case, since the occupation of habitable space is continuous, a daily resolution for the hydraulic model is considered adequate. This temporal resolution may be adjusted according to the variability of discharge in the study reach.

In accordance with the physical habitat template concept (Poff and Ward 1990), the total duration of the simulation should encompass several generations of the target organism. For univoltine species, for instance, four generations would correspond to four years of simulated discharge. This is significantly longer than in common river engineering applications (e.g., flood routing, structure design, etc.).

However, it must be recognized that any improvement on the level of detail with which hydromorphology is described will be associated with an increase in the analysis effort. Fortunately, despite their increased computational cost, simulations of this length are technically feasible today, even with personal computers.

Naturally, even longer simulation times (decades) increase the chances of capturing extreme events in the analysis, which may enhance the description of hydromorphology proposed here. Unfortunately, model run time may become prohibitive unless high-performance computing is available. For hydro- and morphodynamic simulations of this order, simplified grid-based 2D models common in fluvial geomorphology may be used (e.g., Bates et al. 2010, Coulthard et al. 2007), although, careful consideration must be given to model resolution and process representation.

7.2.2 Limitations in process representation

The current state of the art in river hydraulic simulation is such that much of the small-scale complexity of natural rivers is commonly not represented by existing tools. In particular, near-bed, small-scale substrate and flow structures are 'averaged out' by the model due to the prohibitive computational cost of the mesh resolution required to model them explicitly.

However, drag forces, turbulence and mixing are among the most salient characteristics of the boundary layers inhabited by lotic benthos (Leclerc 2005); therefore, existing models must continue to be developed if they are to be used reliably in analyses of the structural quality of these environments.

Based on these concerns, a new trend has developed in river hydraulics in the recent years whose aim is, among others, to provide hydraulic models capable of representing the hydrodynamic complexity of natural rivers. This trend has been grouped by Nikora (2010) under the name of 'hydrodynamics of aquatic ecosystems' (Nikora 2010), who has also identified several possible research directions within this new field.

At the same time, different lab flume studies have been conducted in order to estimate vertical velocity distributions over rough beds (Baptist 2005, Bezzola 2002, Nikora et al. 2007). These have resulted in a series of semi-empirical approaches that allow linking depth-average and near-bed flow-velocities in open channels.

In the near future, 2D codes making use of such approaches may become more widespread in eco-hydraulics, which will allow 'adjusting' depth-averaged velocities from shallow water models and provide better estimates of flow conditions near the bed.

In the present approach, shallow water simulations were adjusted using Bezzola's (2002) method, which is based on an adjustment of the vertical distribution of turbulent shear stress (τ_t) for hydraulically rough beds.

Hydraulic roughness is defined according to the relative submergence of the flow, i.e., the ratio between the flow depth (h) and a representative measure of the size of the roughness elements, k. According to Bray (1987) (cited in Bezzola 2002), flow with a relative submergence below a value of 5.3 can be considered rough.

Under these conditions, the vertical distribution of turbulent shear stress can be approximated by the function shown in figure 7.2. Secondary flows induced by roughness elements lead to the development of a *roughness sublayer*, whose thickness can be estimated based on the size of the roughness elements relative to water depth (Bezzola 2002). Above this layer, turbulent shear stress makes up most of total shear stress (τ) (i.e., viscous effects are negligible), which increases linearly with depth.

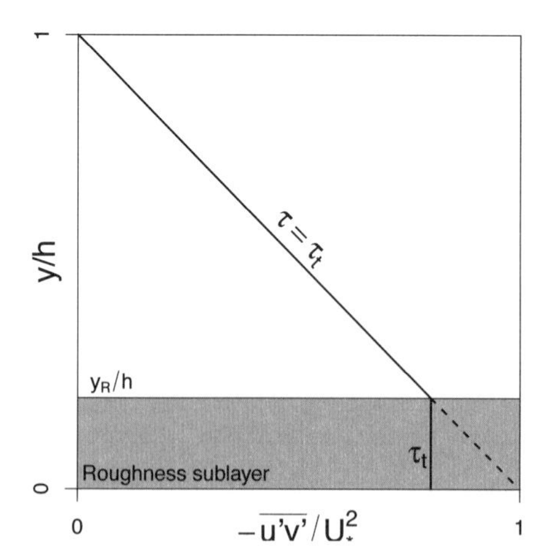

Figure 7.2: Approximated vertical distribution of turbulent (τ_t) and total (τ) shear stress for flow over rough beds, redrawn from Bezzola (2002). y/h = relative submergence, y = vertical position, h = water depth, y_R = thickness of the roughness sublayer, U_* = shear velocity, $\overline{u'v'}$ = turbulent shear stress per unit mass in the vertical-streamwise direction.

Within the roughness sublayer, secondary flows induced by roughness elements enhance mixing and dampen turbulence intensity ($\overline{u'u'}$, $\overline{v'v'}$, $\overline{w'w'}$), causing turbulent shear stress to remain approximately constant from $y = y_R$ to $y = 0$ (y_R = thickness of the roughness sublayer). This phenomenon is accounted for by means of the damping factor ${c_R}^2$:

$$
{c_R}^2 = \begin{cases} 1 - \frac{h}{y_R} & \frac{h}{y_R} > 2 \\ 0.25\frac{h}{y_R} & 0 \leq \frac{h}{y_R} \leq 2 \end{cases} \tag{7.1}
$$

which is used in the proposed model to adjust the vertical velocity profile according to the aforementioned processes. The resulting equations for the (dimensionless) vertical velocity distribution are (Bezzola 2002, equations 8.34a and b):

$$
\frac{\overline{u}}{U_*} = \begin{cases} c_R\left(\frac{1}{\kappa}ln\frac{y}{y_R} + 8.48\right) & y \leq y_W \\ c_R\left(\frac{1}{\kappa}ln\frac{y_W}{y_R} + 8.48\right) + \frac{2}{3}\frac{h}{\kappa y_W}\left[\left(1 - \frac{y_W}{h}\right)^{\frac{3}{2}} - \left(1 - \frac{y}{h}\right)^{\frac{3}{2}}\right] & y > y_W \end{cases} \tag{7.2}
$$

where $\overline{u}$ = mean velocity along main flow axis [m/s], U_* = shear velocity [m/s], c_R = damping factor [-], κ = von Kármán constant (0.41), y = vertical coordinate [m], y_R = thickness of roughness sublayer [m], y_W = thickness of inner region [m] and h = water depth [m]. Figure 7.3 illustrates these quantities.

7.3 SUBSTRATE MAP

Substrate maps can be obtained by different methods, depending on characteristics of the study reach such as width, depth, flow velocity, water turbidity, riparian vegetation cover and substrate complexity, as well as on available resources.

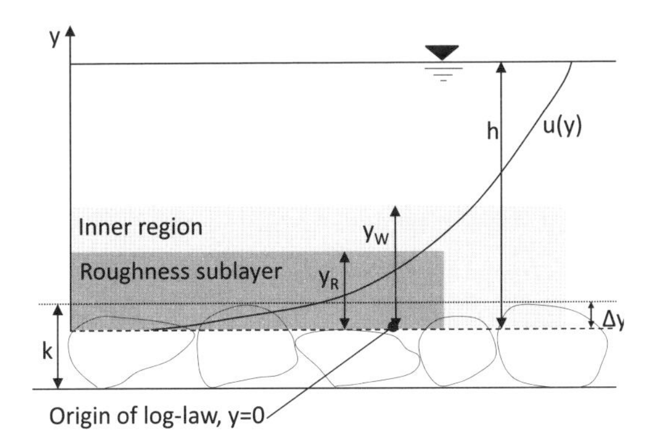

Figure 7.3: Schematic representation of the quantities in Bezzola's velocity distribution model. y = vertical coordinate, $u(y)$ = vertical distribution of stream-wise velocity component, k = geometric roughness height, y_R = thickness of the roughness sublayer, y_W = thickness of the inner (wall) region, h = water depth, Δy = zero-plane displacement. Redrawn from Bezzola (2002).

Manual methods of substrate mapping include bottom sampling and direct observation of substrate (either with the naked eye or through videography), accompanied by analog or electronic positioning techniques. However, automated methods based on hydro-acoustic technology have recently become available (e.g., Reuter et al. 2008, Shields Jr. 2010).

These methods employ the 'return signal strength from the bottom track echo and use this information to compute the echo intensity EI at the bed' (Shields Jr. 2010). Distinct signals can thus be recognized for the different types of bottom cover present in a reach; however, careful classification analyses are required to relate the statistical properties of the return signal to the different substrate types, and clear distinctions between acoustically similar substrates may not always be possible (Reuter et al. 2008, Shields Jr. 2010).

Remote sensing of instream and floodplain habitats is also possible as long as water turbidity or riparian vegetation do not block the view of the bottom (Marcus and Fonstad 2008). Different alternatives exist within this group of approaches. Sensors mounted on aircraft, satellites or even drones may be used, depending on the extent and resolution of the study. For a summary of available techniques, see Marcus and Fonstad (2008) and references therein.

Substrate maps are a crucial input to the approach presented here. Due to the small scale at which resource patches and organism activity occur for benthic invertebrates, their resolution must be below the meter level, which is within the range achievable with available mapping methods.

Ideally, the resolution of the substrate map must be between 2 and 5 times smaller than the smallest substrate patch, as has been proposed in spatial ecology studies (O'Neill et al. 1996). However, in the common situation that only punctual substrate data (e.g., grab-samples) are available, the alternative approach presented in section 6.2.4.3 can be followed.

Different schemes may be used for substrate classification. However, the choriotop categories according to Austrian Standard OEN M6232 (ÖNORM-M-6232 1997) have been specifically designed for benthic habitat classification, and are therefore a suitable alternative (table 7.1).

Given that the modeling time for this approach must encompass several generations of the target species, substrate maps may have to be updated in the course of the simulation, particularly if

Table 7.1: Choriotop classification categories according to Austrian Standard OEN M6232 (ÖNORM-M-6232 1997).

Substrate type	Description
Pelal	Silt, loam, clay and sludge (<0.063 mm)
Psammal	Sand (0.063 - 2 mm)
Akal	Medium to fine gravel (0.2 - 2 cm)
Microlithal	Egg to fist size cobbles with mixture of medium to fine gravel (2 - 6.3 cm)
Mesolithal	Fist to hand size cobbles with a mixture of medium to fine gravel (6.3 - 20 cm)
Macrolithal	Hand to head size boulders, gravel and sand (20 - 40 cm)
Megalithal	Large cobbles, blocks, and bedrock (>40 cm)
Gigalithal	Bedrock
Sapropel	Organic sludge
Detritus	Deposits of coarse (CPOM) or fine (FPOM) particulate organic matter
Debris	Organic and inorganic matter deposited within the splash zone area by wave motion and changing water levels (e.g. mussel and snail shells)
Phytal	Submerged plants, floating stands or mats, lawns of bacteria or fungi, and tufts, often with aggregations of detritus, moss or algal mats. (Interphytal = habitat within a vegetation stand or plant mat)
Xylal	Tree trunks, roots, branches or other dead wood

there are large discharge events able to change the distribution of substrates and the morphology of the study reach.

In this case, morphodynamic simulations may have to be coupled to the hydrodynamic model in order for the analyses to remain as realistic as possible.

7.4 HABITAT SUITABILITY ASSIGNMENT

As discussed in section 6.2.3, habitable space is designated based on the overlay of substrate, velocity and water depth data at each time step of the habitable space simulation.

These data enter the analysis as either pixel maps (i.e., rasters) or point maps depending on whether the substrate information available is continuous or punctual. In this context, pixels correspond to the resolution of the hydrodynamic model and substrate map, and, consequently, of the habitable space maps resulting from their overlay process.

This overlay can be done through different approaches, all of which are based either on empirical data or expert knowledge on the target species (Leclerc 2005). Several authors have summarized the advantages and restrictions of these approaches. For detailed discussions on this, see for instance Caissie and El-Jabi (2003), Capra et al. (1995), Didderen and Verdonschot (2010), Leclerc (2005), Parasiewicz (2007).

In technical terms, the overlay can be done using the scripting capabilities of any GIS software. In this thesis, R was used (see script in chapter 11). The algorithm required for this task is a simple loop that iterates through all time steps, performing each time an overlay operation combined with a conditional statement that designates the pixel (or point, see below) as habitable or inhabitable according to the species' tolerance.

This task is done on a pixel-by-pixel or point-by-point basis depending on whether punctual or continuous (e.g., raster) substrate data are available. A new matrix or xy point file of habitable space is written out at each time step.

These files are then used as input for the patch building process.

7.5 PATCH BUILDING

The next step in describing the spatiotemporal distribution of habitable space consists in building habitable space patches out of adjacent habitable pixels at each time step in the analysis. This process cannot be done in cases where no continuous substrate data are available, for which the alternative presented in section 6.2.4.3 shoud be used.

Although early proposals for spatiotemporal analyses of this sort in stream ecology date back more than two decades (Pringle et al. 1988), their technical implementation has, to the knowledge of the author, only become widespread in the terrestrial metapopulation ecology (e.g., Akçakaya 2000, Hanski 1999b).

At each time step, the patch-building process assigns a patch ID (patch identification) number to habitable cells, ignoring those that are inhabitable. Habitable cells that are within a specified distance of each other will be assigned the same patch ID, and therefore will be categorized as belonging to the same patch.

The function used for patch ID assignment in the present approach is the 'clump' function of R package 'raster' (Hijmans 2013), which clumps together adjacent cells within their Moore neighborhood (8 neighbors; north, south, east, west and diagonals).

An illustration of how patches are built following this approach is presented in figure 7.4. The corresponding script for this task is presented in chapter 11.

The result of the patch-building process can be written out as a raster or polygon file at each time step. These files can then be used as input in the patch tracking proess.

7.6 PATCH TRACKING ALGORITHM

The outcome of the patch-building process is a set of patch maps, one for each time step in the hydraulic simulation. In its present form, the algorithm stores patch maps in ESRI's polygon shapefile format. This helps reducing requirements regarding computer memory and processing capacity relative to what would be needed in raster format.

With this set of patch maps, a dynamic description of habitable space can be generated, which can be visualized by plotting the sequence of maps from all time steps in the analysis period. Animations can be readily made in R using package 'animation' (Xie 2013).

This type of visualization reveals the magnitude of variability in habitable space, something that is not possible with static approaches and which is recognized in ecological theory as an important determinant of habitat quality (Southwood 1977).

The next step constitutes one of the key technical aspects of this approach. Given that patches may appear, disappear, expand or contract throughout the simulation, an algorithm was developed to track the behavior of individual patches automatically throughout the analysis period.

For this, patch maps are read in sequentially one time step at a time, each patch present at time *t* is assigned a unique **cell signature** (see below), and cell signatures are compared among all

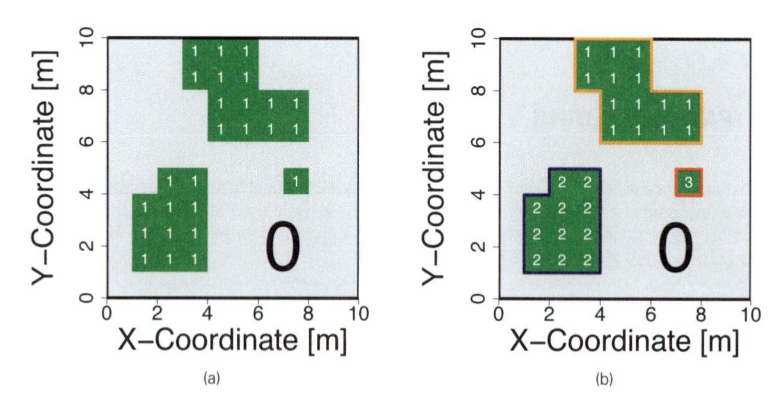

Figure 7.4: Hypothetical data used to illustrate the patch building process. (a) Suitable pixels. (b) All adjacent pixels within the Moore neighborhood of a cell have been assigned the same patch ID value, such that all cells belonging to the same patch share the same ID code (seen here color-coded). All pixels in the grey area have a value of zero, i.e., are inhabitable.

patches and time steps in order to build an list of all the signatures of all the patches that existed in the modeled reach during the analysis period.

The cell signature of a patch is simply a vector of integer numbers obtained by overlaying the patch map onto a numbered grid of the same extent and resolution as the substrate raster (figure 7.5). Pixels in this numbered grid are numbered from 1 to n, with n = number of pixels in the substrate raster. No repetition of cell numbers is allowed, which provides a unique cell signature for each polygon in the patch map.

In the example of figure 7.5, three patches exist at time step t_1 (a), with the cell signatures <4, 5, 6, 14, 15, 16, 25, 26, 27, 28, 35, 36, 37, 38>, <53, 54, 62, 63, 64, 72, 73, 74, 82, 83, 84> and <58>. At the next time step (b), the first patch in (a) has not changed, the second has the reduced cell signature <53, 62, 63, 72, 73, 82, 83>, the third has the increased signature <57, 58, 67, 68>, and there is a new patch with the signature <86, 87>.

With these unique cell signatures it is possible to track exactly the behavior of individual patches between consecutive time steps. If the size of a patch increases between t and $t+1$, new cells will be added to its cell signature; if it becomes smaller, its cell signature will become shorter (it may also disappear completely, in which case its cell signature will be empty and patch area will be set to zero).

If a patch disappears and re-appears at a later time, it will be interpreted as the same patch and its area will be written to the same column of the areas matrix. If a patch with a completely new

Table 7.2: Hypothetical areas matrix showing the areas of three patches at four time steps.

Time step	**patch1**	**patch2**	**patch3**	**...**
1	1.31	0	4.51	...
2	0.42	0.16	2.11	...
3	1.37	0	1.18	...
4	2.56	0.14	0.11	...
...	...	...	...	...

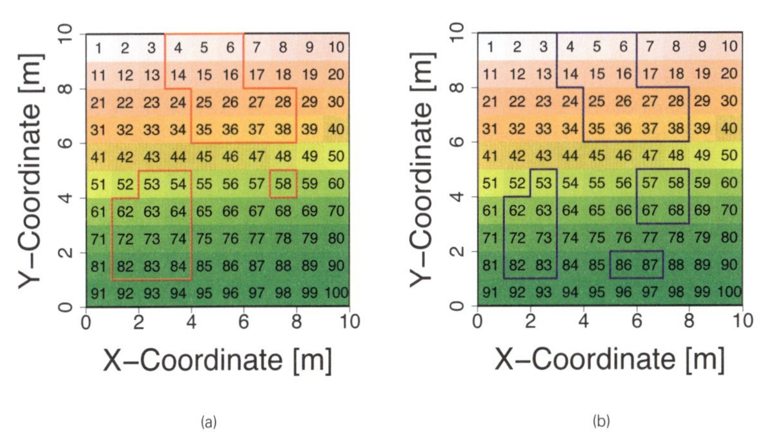

(a) (b)

Figure 7.5: Illustration of the patch tracking algorithm proposed in this thesis using a hypothetical patch network as an example. (a) Patch network at time step t_1 with three patches (red polygons). (b) The blue polygons indicate the patch network at the next time step t_2.

cell signature appears, it is interpreted as a new patch altogether and a new column will be added to the areas matrix.

It is important to note that the comparison of patch cell signatures is based not only on their length but also on the ID values of the cells making up the patches. A cell signature is considered new when no common cells exist with any other signature in any time step. Conversely, cell signatures that contain at least one common cell are interpreted as belonging to the same patch.

Multiplying the length of each patch signature at each time step by the pixel size returns the area of each patch in the network at each time step. This result is referred to here as the **areas matrix** (table 7.2).

This matrix represents a valuable source of ecological information. It describes the dynamics of each individual patch of habitable space in the study reach in terms of its changes in size due to hydrodynamic and substrate variations.

The patch tracking algorithm is presented as an R script in chapter 11.

With this information, the patch dynamics indicators proposed in section 6.2 can be calculated.

7.7 CALCULATION OF PATCH DYNAMICS INDICATORS

The calculation of the patch dynamics indicators described in can be automated using R scripts. A first version of these scripts is presented in chapter 11.

Indicators are grouped as in section 6.2, i.e., into indicators of patch size and size distribution, patch duration and patch juxtaposition. All computations are based on the areas matrix, an associated centroids matrix, the patch maps and the 2D velocity field resulting from the hydrodynamic model.

The first set of indicators, which relates to patch area and duration analysis, is calculated by script 'ind_area_analysis.R'. In it, the areas and centroids matrices are read in and used to calculate spatially-explicit, patch-specific area differences, as described in section 6.2.

The output of this step comprises time series for the number of patches, the total habitable area, patch-specific area losses l_i (as an R list object) and l_Δ. All these indicators are stored in separate text files in the folder '\output'.

The second set of indicators results from execution of script 'ind_patch_duration.R'. Using the same input data, this script produces text files containing the start and duration of all patch-specific existence spells (as an R list object), the duration of inter-spell periods (also as an R list object), and patch-specific, spatially explicit values for total, minimum and average patch duration, number of existence spells, and average, minimum and maximum inter-spell duration.

The analysis of patch juxtaposition is performed by script 'ind_journeyTrack.R'. The output of this method is stored in a text file containing a time series for *c*.

In the current version of the scripts, the spatial distribution of variables is built using the time average of each patch's centroid. The temporal resolution for computation of all indicators can be controlled by modifying the header of the R for loop in the corresponding scripts.

A guide to all scripts and associated files is presented in chapter 11.

8 PROOF OF CONCEPT

8.1 INTRODUCTION

The methodological framework presented in this thesis provides an alternative perspective for looking at river hydromorphology and its relationship to ecological quality. It must therefore be based on a link between the habitat dynamics indicators derived in the method and the biological populations of interest.

In this chapter, an empirical check, or proof-of-concept, of the existence of such a link is conducted using available macrozoobenthos and hydromorphological data, which were kindly provided by the University of Duisburg-Essen's Aquatic Ecology Department (Januschke et al. 2014, Jähnig et al. 2008).

It is important to mention that the theoretical basis of the method (Hart and Finelli 1999, Poff and Ward 1990, Pringle et al. 1988, Southwood 1977, Statzner et al. 1988, Townsend 1989, Townsend and Hildrew 1976, among others) derives from decades of observational and experimental studies in ecology.

This means that the principles underlying the approach are themselves the result of empirical studies rather than theoretical conjecture. Thus, the analyses presented here can be understood as a confirmation of the applicability of these principles to a particular test case.

Key to this proof-of-concept is that the sampling sites differ only with respect to their hydromorphology, or, more specifically, with respect to the spatiotemporal dynamics of aquatic habitable space, defined here as space that is aquatic, i.e., not dry, and mechanically suitable for habitation at a specific point in time and space given its topography, substrate, discharge and water depth.

The remaining four factor groups (food sources, water quality, source populations, and biotic interactions; Schuwirth 2012) must be within comparable ranges, otherwise any relationship found between the habitat dynamics indicators and the macrozoobenthos community could not be attributed to hydromorphology.

The link discussed here relates biological populations and only one of the five factor groups influencing riverine macrozoobenthos (see section 5.1). For this reason, the type of relationship to be expected should not be of central-response but rather limiting-response type (*sensu* Lancaster and Belyea 2006).

As discussed in section 5.1, the latter are more realistic in an ecological context, where response variables are influenced by a multiplicity of factors simultaneously, including different forms of stochasticity (Shaffer 1981) and spatial dynamics (Downes and Lancaster 2010).

The basis of the test presented here is a planned comparison (*sensu* Downes 2010) between the average local density of ecologically different species in contrasting hydromophological settings, namely trained and morphologically restored sites.

The ecological differences among species are based on their sensitivity to losses of aquatic habitable space. Expert knowledge used for this is summarized in a macroinvertebrate trait database compiled at the German Federal Institute of Hydrology's (Bundesanstalt für Gewässerkunde) Animal Ecology Department, which was kindly provided in cooperation for this thesis.

Lastly, it is important to mention that the analyses in this chapter represent the application of the proposed framework in a praxis-related scenario, i.e., using the type of data that are commonly available in river management praxis (at least in Germany). Hence, this proof-of-concept can also be seen as a check of the applicability of the method.

8.2 METHODOLOGY

8.2.1 Available data and study area

The base data used for this test application come from a comparative study of trained and restored reaches of the upper river Lahn (federal state of Hessen, Germany) carried out by Januschke et al. (2014) and Jähnig et al. (2008) (figure 8.1).

According to the German river typology (Pottgiesser and Sommerhäuser 2008), the Lahn is a "mid-sized fine to coarse substrate dominated siliceous highland river" (type 9). This type is characterized by floodplain widths on the order of 200-300 m, a slightly meandering-branched pattern, and substrate dominated by cobbles and rocks with interspersed deposits of finer sediments in slow-flowing areas.

Channels are typically wide and shallow, and flow is predominantly fast and turbulent with localized slow-flow areas near the banks, backwater areas and behind large rocks. The slope of the valley floor is typically between 0.2% and 0.6%.

Januschke et al. (2014) compared the macrozoobenthos community in three trained and three restored sites on the Lahn with the objective of exploring the effects of restoration on this organism group.

Table 8.1: Geographic position, catchment area, altitude, restoration year and sampling dates (1 = June 2005, 2 = July 2007, 3 = June 2009, 4 = June 2012) for the three study sites sampled by Januschke et al. (2014) and Jähnig et al. (2008). Table reproduced after Januschke et al. (2014) and Jähnig et al. (2008).

Site	**Cölbe**	**Ludwigshütte**	**Wallau**
Latitude (N)	50° 51'47"	50° 55'29"	50° 55'37"
Longitude (E)	8° 47'25"	8° 29"59"	8° 29'20"
Catchment size (km^2)	650	288	278
Altitude (m. asl.)	190	300	300
Mean discharge (m^3/s)	8.3	5.2	5.1
Local channel slope (m/km)	0.20	0.40	0.21
Restoration year	2000	2002	2001
Sampling dates	1,2,3,4	1, 3	1,3

At each of three localities (Cölbe, Ludwigshütte and Wallau), a restored and a nearby upstream trained site were sampled following a multi-habitat sampling technique (Meier et al. 2006), with 20 sample units per site and substrate types sampled in proportion to their coverage (Januschke et al. 2014).

Table 8.1 summarizes the general characteristics of each site.

Next to the biological samples, abiotic data were also collected at the study sites (Jähnig et al. 2008) which are relevant to this test application. These included punctual substrate and water depth observations at all sampling sites and dates, each time using an array of 100 points distributed equally in 10 cross sections 20 meters apart (i.e., 10 points x 10 cross sections) in each site.

Substrate information was available in the form of 100 punctual observations at each locality, which means that the habitable space availability indicator used was **hsAv** (see section 6.2.4.3).

Water depth data were used in combination with survey profiles (kindly provided by the Hessisches Landesamt für Umwelt und Geologie) and aerial ortophotography (acquired from the Hessische Verwaltung für Bodenmanagement und Geoinformation) in order to build an elevation model of each of the six (3 restored, 3 trained) study reaches. Figure 8.2 shows aerial ortophotos of the six study sites.

In addition, discharge time series were obtained (kindly provided by the Hessisches Landesamt für Umwelt und Geologie) for gauges on the Lahn (gauges Sarnau and Biedenkopf) and a tributary that discharges approximately 800 m upstream of site Cölbe (the Wetschaft, MQ = 1.701 m^3/s, gauge Niederwetter).

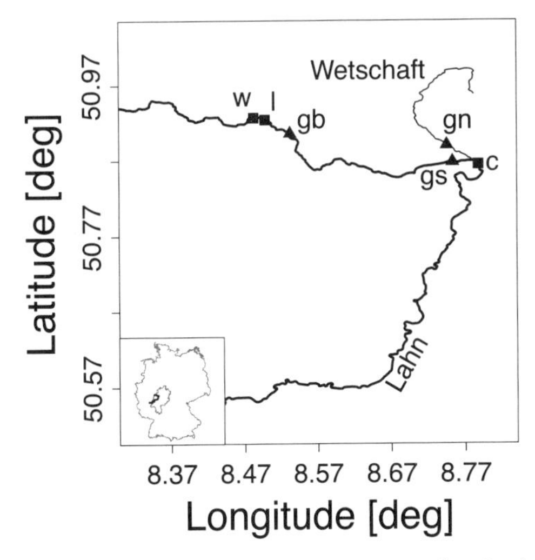

Figure 8.1: General overview of the study area. Squares indicate sampling sites (w = Wallau, l = Ludwigshütte and c = Cölbe); triangles indicate gauging stations on the Lahn (gb = gauge Biedekopf, gs = gauge Sarnau) and a tributary (the Wetschaft, gn = gauge Niederwetter). The inset shows the location of the Lahn river in Germany and the Federal State of Hessen. Drawn using data from Januschke et al. (2014), ©GeoBasis-DE/BKG 2015, the Hessische Verwaltung für Bodenmanagement und Geoinformation and the Hessisches Landesamt für Umwelt und Geologie.

These time series are presented in figures 8.3 (b) to (d). Gaps in the discharge time series of gauge Sarnau, which covered about 23% of the observation time, were filled by linear correlation with the time series of gauge Biedenkopf, located 23 km upstream. This correlation was built after removing outliers, defined here as points with a residual of more than 30 m^3/s with respect to the ordinary least squares regression line (see figure 8.3a).

8.2.2 Analysis design and research hypothesis

The objective of this analysis is to explore the relationship between habitable space availability, as summarized by hsAv, and the average local density (*sensu* Lancaster and Belyea 2006) of selected species.

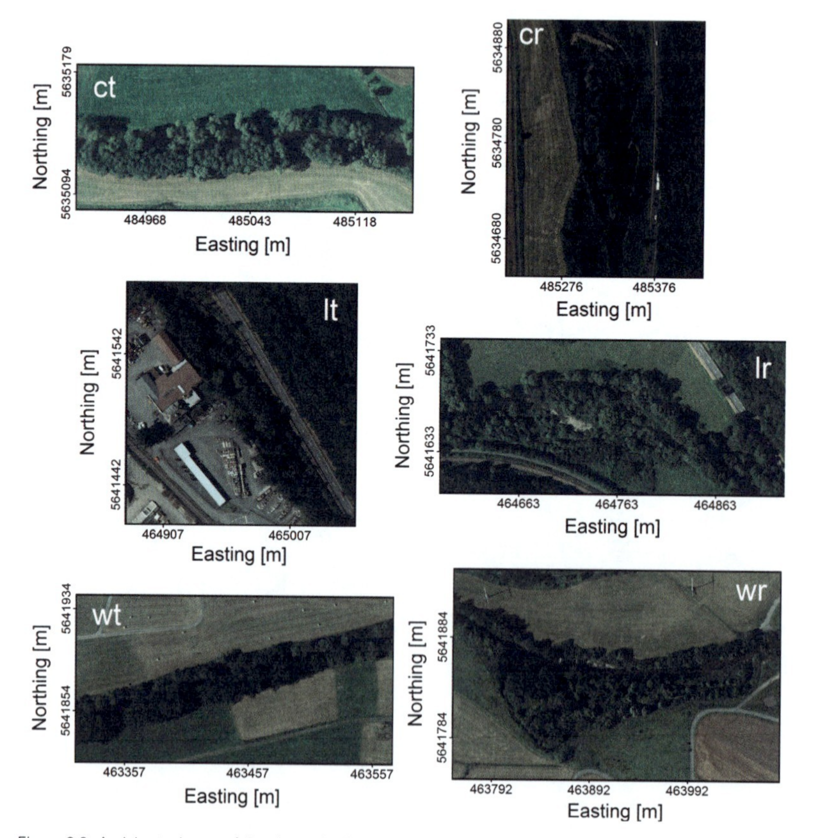

Figure 8.2: Aerial ortophotos of the six study sites. ct = Cölbe trained, cr = Cölbe restored, lt = Ludwigshütte trained, lr = Ludwigshütte restored, wt = Wallau trained, wr = Wallau restored. Images acquired from the Hessische Verwaltung für Bodenmanagement und Geoinformation. Projected coordinate system: ETRS89-UTM32, unit: meter.

Average local density (**avgN**) was calculated here as the mean value of the density records of a species at a single site. Site names were coded as follows: ct = Cölbe trained, cr = Cölbe restored, lt = Ludwigshütte trained, lr = Ludwigshütte restored, wt = Wallau trained, wr = Wallau restored.

Exploring the aforementioned relationship allows checking whether a reduced supply of aquatic habitable space, in terms of availability and duration of habitable area, poses a limit to the local density of the analyzed taxa.

The use of a limiting response (LR) framework is a more realistic approach in this type of ecological analysis (Lancaster and Belyea 2006) given that response variables can be influenced by a large number of factors, only some of which are involved in the analysis (see section 5.1).

Samples from all sites were pooled together for these analyses. Sites ct and cr had 4 samples, while all other sites had 2. The spatial (> 500 m) and temporal (> 2 years) separation between

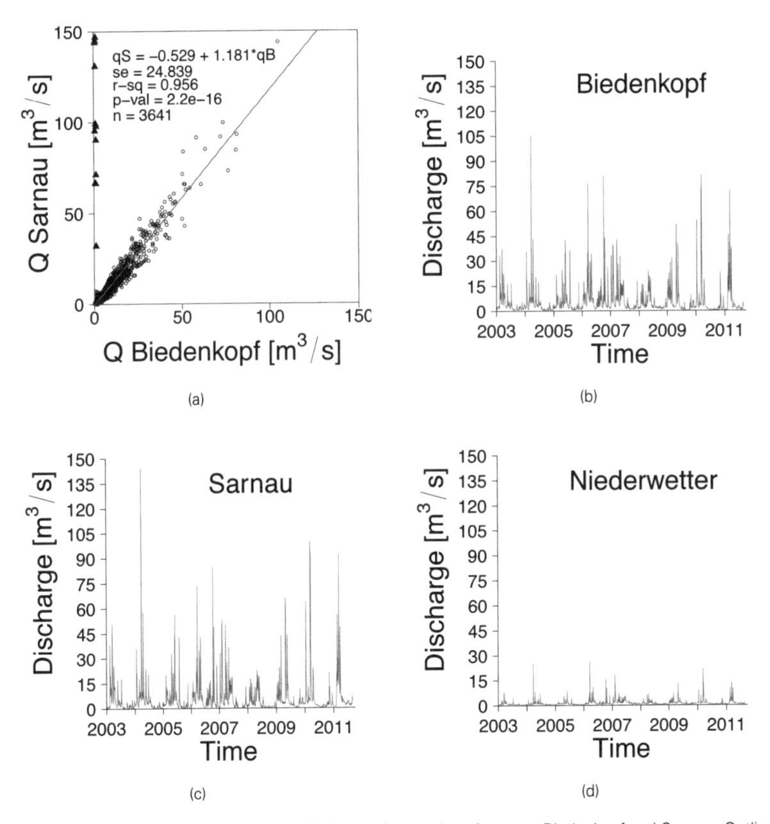

Figure 8.3: (a) Relationship between the discharge time series of gauges Biedenkopf and Sarnau. Outliers are shown as triangles. These points have a residual of more than 30 m^3/s with respect to the ordinary least squares regression line (grey line). (b) to (d) Discharge time series for gauges Biedenkopf, Sarnau and Niederwetter, respectively. For symbols in plot, see List of Symbols.

samples is enough that they can be considered independent with respect to their biological information (each sample captured a separate, independent biological population or sub-population).

The average local density (ind./m^2) of species of different degrees of sensitivity was explored along a gradient of hsAv. Species sensitivity is understood here as the lack of traits that may confer resistance to habitat loss, which include high mobility, numerous offspring, multivoltinism, streamlined body, and long-lived adults, among others (see section 8.2.3).

These analyses were complemented with comparisons between habitat availability indicators between restored and trained sites, which were done in order to check the capacity of the method to distinguish between these two hydromorphological situations.

The hypothesized relationship is the following;

> *With minimal influence of other factors, the availability of habitable space and its spatiotemporal variability set an upper limit to the population success (measured as average local density) of (type-specific) sensitive taxa, i.e., taxa that depend more strongly on a continuous supply of habitable space. The success of tolerant taxa is expected to exhibit no such limitation along the same gradient of habitat availability.*

If the above hypothesis can be confirmed, there will be evidence of a relationship between habitat dynamics and ecological quality, understood in terms of the population success of the target species. In that case, it would be possible to use the proposed indicators to make an objective judgment of local hydromorphological quality.

The degree of certainty of this conclusion grows with the number of tested species and contrasting site pairs (trained vs. restored), as the likelihood that unmeasured factors lead to the observed relationship becomes smaller with increasing number of these design elements (Downes 2010, Keough et al. 1993).

However, it must be recognized that some level of speculation is inherent to any comparative study of this sort (Townsend 1989, Underwood 1991, 1992, 1994, 1989, 1993).

Experimental units in this study correspond to sites*species, i.e., there is a value of average local density for each species at each site. This average density was calculated as the mean of the local densities recorded in each sampling event at each site.

For each of these average densities there was an associated value of the habitat dynamics indicator derived from the proposed method (see section 8.2.6).

It is important to mention that any trait-based ecological analysis will have uncertainty derived from the fact that traits are not independent, and even though a taxon may be conferred resistance by some traits, others may make it more susceptible to fluctuations in habitable space availability (Verberk et al. 2013).

8.2.3 Classification of species sensitivity

The tolerance of the selected MZB species to habitable space scarcity was represented through a resistance value, **resVal**. This index was computed using the information available in a trait database compiled by the German Federal Institute of Hydrology's (Bundesanstalt für Gewässerkunde, BfG) Animal Ecology Department, which was kindly provided in cooperation for this thesis.

This trait database is described in Haybach et al. (2004), and is referred to in the remainder of this chapter as the 'BfG trait database'.

The resistance value (resVal) of a species was computed as follows. First, traits known to increase population success in a variable environment (e.g., long-lived adults, streamlined body form of larvae, numerous offspring, many reproductive cycles, high dissemination potential, etc., among others) were selected (see table 8.2).

Then, each trait category was assigned a rank that ranges from 1 up (linearly) to the maximum number of categories in that trait, with higher ranks being assigned to the categories known to increase population success in a variable environment.

So, for instance, if a trait has 4 categories, the ranks 1, 2 3, and 4 will be distributed among the categories according to the degree of resistance each of them confers.

In the BfG database, the affinity of a species for a trait category is coded from 0 to 1, with 0 = no affinity and 1 = high affinity. The sum of the affinities across all categories of a trait is equal to 1, i.e., affinity within a trait is distributed among its categories.

Table 8.2: Species traits, categories and their rank with respect to resistance to habitable space scarcity. Trait information derived from the trait database of the German Federal Institute of Hydrology's (Bundesanstalt für Gewässerkunde) Animal Ecology Department. Data kindly provided cooperation for this thesis. For a more detailed description of this data base, see Haybach et al. (2004).

Trait	Categories	Rank
Size	<=5 mm	5
	>5-10 mm	4
	>10-20 mm	3
	>20-40 mm	2
	>40 mm	1
Number of descendants per reproductive cycle	<=100	1
	>100-1000	2
	>1000-3000	3
	>3000	4
Number of reproductive cycles per year	semivoltine	1
	univoltine	2
	plurivoltine	3
Number of reproductive cycles per individual	1	1
	2	2
	>2	3
Life duration of adults	<=1 day	1
	>1-10 days	2
	>10-30 days	3
	>30-365 days	4
	>365 days	5
Reproductive technique	single individual	3
	hermaphrodism	2
	male and female	1
Dissemination potential in water	<=10 m	1
	>10-100 m	2
	>100-1000 m	3
	>1000-10000 m	4
	>10000 m with ships as vector	5
Body flexibility	none (<10°)	1
	low (>10-45°)	2
	high (>45°)	3
Body form of aquatic stages	streamlined	4
	flattened	3
	cylindrical	2
	spherical	1

For each category in a trait, category rank and species affinity for that category were multiplied, and the resulting values were summed to produce a total score for the trait. This score was standardized by dividing it by the maximum possible score for that trait, leading to a number between 0 and 1.

Finally, these trait scores were summed across all traits and subsequently divided by the number of traits (9), such that the final resistance value (resVal) is a number between 0 and 1. With this classification, species with a high affinity for resistance-conferring trait categories will have a high resistance value (closer to 1) and vice versa (closer to zero).

This simple approach (compare for instance with Verberk et al. 2013) attempts to summarize existing biological knowledge into a quantitative index that can be used as covariate in the statistical analyses. Traits and their qualifications are shown in table 8.2. Table 8.3 shows the species selected for these analyses and their resistance value (resVal).

8.2.4 Hydrodynamic numerical modeling of the sampling reaches

A two-dimensional hydrodynamic numerical model was set up and calibrated for each of the six sampling reaches (figure 8.2). The objective of these simulations was to provide a picture of the flow patterns at each site and their spatiotemporal variability at the scales discussed in section 7.2.

Following these principles, the spatial resolution of all models was of 1 m x 1 m, extending over a river length of approximately 200 m. The width of the modeled region was adjusted following the local characteristics of each reach. For trained sections it was between 40 m and 70 m, and for restored sections between 80 m and 100 m.

The output temporal resolution was 1 day in all models, and the extent varied between 5.65 years for sites Wallau and Ludwigshütte and 8.66 years for Cölbe.

These temporal extents were chosen with the purpose of covering at least two hydrological years (the hydrological year in Germany spans from November 1st to October 31) before the first biological sampling, and extended until the day of the last sampling event.

Table 8.3: Species used in the analysis and their resistance values (resVal). For a description of resVal's computation, see text of this section.

Order	Species	Resistance value (resVal)
Ephemeroptera	*Baetis fuscatus*	0.649
	Baetis lutheri	0.637
	Caenis luctuosa	0.597
	Ecdyonurus venosus	0.564
	Ephemera danica	0.508
	Habrophlebia lauta	0.562
	Potamanthus luteus	0.516
Isopoda	*Asellus aquaticus*	0.702
Trichoptera	*Athripsodes albifrons*	0.497
	Athripsodes cinereus	0.480
	Goera pilosa	0.574
	Odontocerum albicorne	0.472
	Psychomyia pusilla	0.606
	Allogamus auricollis	0.498
	Anabolia nervosa	0.506

8.2.4.1 Model geometry

Model geometry is shown in figure 8.4. These elevation models were built combining available information from different sources. The primary source were the water depth data collected by Januschke et al. (2014).

These authors used a grid of 100 points distributed equally in 10 cross sections 20 meters apart (i.e., 10 points x 10 cross sections) at each site. The distribution of these points is shown in figure 8.7.

For these data, the water surface was used as a local datum for all elevations points at each site. An R script was developed (see section 11) in order to perform a linear interpolation along arcs drawn manually using the available aerial pictures as a visual guide.

Interpolated points were then converted into a regular 1 m x 1 m grid using the functions of R's raster package (Hijmans 2013). The resulting elevation models were then tilted downstream using the average longitudinal river slope, which was derived using available cross sectional profiles (30 m 50 m apart) kindly provided by the Hessisches Landesamt für Umwelt und Geologie.

It is important to mention that the point grid used by Januschke et al. (2014) was only locally referenced, and no high-accuracy GPS equipment was used to geolocate each point as is commonly done in technical topographic and bathymetric surveys.

Hence, the elevation models constructed here describe only the major topographical features of the study reaches at the moment of sampling.

However, as discussed in section 6.2.2, the main purpose of hydrodynamic modeling in this method is to depict the spatiotemporal distribution of the flow patterns to which macrozoobenthos taxa are know to be associated (e.g., 'slow-flowing backwater areas of fine sediment', or 'small rocks and cobbles in fast-flowing areas').

Thus the level of accuracy achieved with the derived elevation models is considered sufficient.

8.2.4.2 Boundary conditions and bottom roughness

The upstream boundary conditions for the models were constructed using daily discharge time series from nearby gauges (kindly provided by the Hessisches Landesamt für Umwelt und Geologie). These data are described in section 8.2.1 and figure 8.3.

For models Cölbe restored and Cölbe trained, the time series shown in figure 8.5a was used, which resulted from adding together the discharges of gauges Sarnau and Niederwetter (see figure 8.1).

For the four remaining models (Ludwigshütte restored, Ludwigshütte trained, Wallau restored and Wallau trained), the time series in figure 8.5b was imposed, corresponding to the discharge time series of gauge Biedenkopf (see figure 8.1).

A constant energy slope was used as downstream boundary condition in all models (table 8.4). These values were assumed equal to the average longitudinal river slope at each study site.

A warm-up period of one day was used at the beginning of each simulation to 'fill' the models with water and avoid instability.

Manning's bottom roughness coefficient *n* was first selected based on the substrate information collected by Januschke et al. (2014), aerial ortophotos acquired from the Hessische Verwaltung für Bodenmanagement und Geoinformation, and Te Chow (1959).

n values were selected separately for the different areas of each site, e.g., vegetated islands,

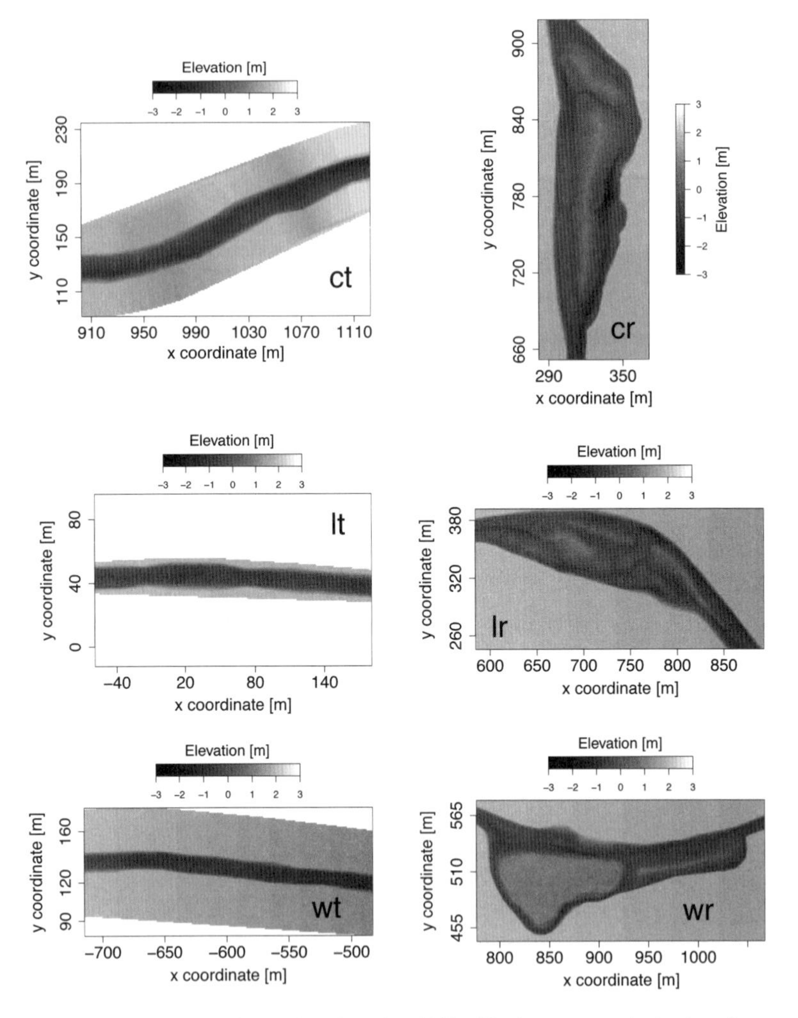

Figure 8.4: Elevation models for the six study reaches. X, Y and Z values correspond to local coordinate systems for each site. ct = Cölbe trained, cr = Cölbe restored, lt = Ludwigshütte trained, lr = Ludwigshütte restored, wt = Wallau trained, wr = Wallau restored. Elevation models constructed using data from Januschke et al. (2014), the Hessisches Landesamt für Umwelt und Geologie and the Hessische Verwaltung für Bodenmanagement und Geoinformation. Data points were rotated to align model boundaries vertically (or horizontally) as required by the hydrodynamic numerical code LISFLOOD-ACC (Bates et al. 2010).

sandy patches, cobble- or gravel covered bottom, etc. These starting n values were then modified in the calibration process (see section 8.2.4.5). The resulting bottom roughness maps are shown in figure 8.6.

8.2.4.3 Hydrodynamic numerical code

One of the guiding principles of the methodological framework presented here is that the temporal extent of the hydraulic simulations must cover several generations of the target species (see sections 6.2.1 and 7.2.1). This is a critical point in terms of the computational cost of the indicators on which the method is based.

Numerical shallow water codes commonly used in hydraulic engineering are typically based on some form of the depth-averaged 3D Reynolds equations, and simulated time spans typically range between a few hours and a few days or weeks.

In contrast, simulating habitat dynamics over several generations of, say, a univoltine macrozoobenthos taxon, would imply modeling over time spans of several years.

The CPU time of such a simulation would well exceed those found in common hydraulic simulation practice, even with modern workstations. Hence, since high-capacity computing is rarely available in management practice, alternative modeling strategies must be sought in order to make the method applicable in this context.

Table 8.4: Downstream boundary condition (energy slope at downstream boundary) of the six models. ct = Cölbe trained, cr = Cölbe restored, lt = Ludwigshütte trained, lr = Ludwigshütte restored, wt = Wallau trained, wr = Wallau restored.

Site	Trained	Restored
Cölbe	-0.00165	-0.00165
Ludwigshütte	-0.00100	-0.00200
Wallau	-0.00405	-0.00205

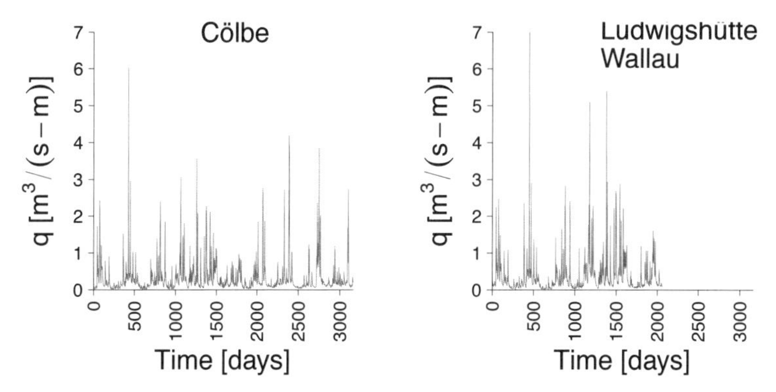

Figure 8.5: Upstream boundary condition in the six models. The curves represent discharge per unit width through the upstream boundary of the model, indicated on the y-axis as q ($m^3/s - m$). The curve for site Cölbe is used for both the trained and the restored locations. The same applies for Wallau and Ludwigshütte.

Given that high-accuracy (e.g., highly-resolved three-dimensional) simulations are not necessary for the method proposed here (see section 6.2.2), a potential strategy for reducing CPU time is the use of models that are based on simplified forms of the flow equations.

This strategy is common in other areas of river science, such as fluvial morphology (Bokulich 2013, Nicholas 2005) and flood risk analysis (Hunter et al. 2007).

These 'reduced complexity models' (Brasington and Richards 2007) have the capacity to reproduce reach-scale hydrodynamic processes with varying degrees of detail depending on the terms of the shallow water equations that are assumed negligible.

Early fluvial morphology models (e.g., Coulthard et al. 1996, Murray and Paola 1994), for instance,

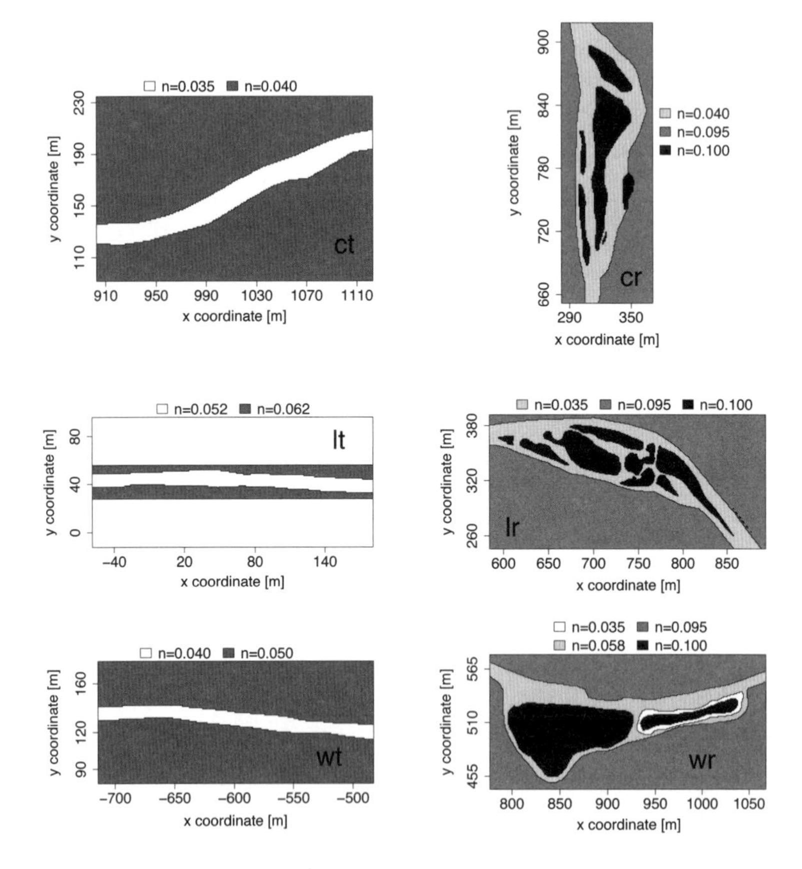

Figure 8.6: Manning roughness maps for the six study reaches. X and Y values correspond to local coordinate systems for each site. ct = Cölbe trained, cr = Cölbe restored, lt = Ludwigshütte trained, lr = Ludwigshütte restored, wt = Wallau trained, wr = Wallau restored. Manning values derived using data collected by Januschke et al. (2014) and suggestions in Te Chow (1959), and further adjusted during model calibration.

use a very simple water routing scheme based on the kinematic wave approximation to the shallow water equations.

Slightly more complex routing algorithms solve the diffusion wave approximation (Bates and Roo 2000), although for both groups issues of mass conservation and stability can arise in practice (Hunter et al. 2007, Nicholas 2009).

8.2.4.4 LISFLOOD-ACC

An alternative simplification proposed by Bates et al. (2010) and further developed by de Almeida et al. (2012), referred to as the 'local inertial approximation', attempts to address these issues with the aim of allowing stable, computationally efficient simulation at large extents and sufficient resolution.

This is the basis of the 2D numerical model used here: LISFLOOD-ACC, which was developed at the University of Bristol's School of Geographical Sciences and is available free of charge for research purposes upon request.

This numerical code is based on the inviscid 2D shallow water equations (de Almeida and Bates 2013):

Mass conservation:

$$\frac{\partial h}{\partial t} + \frac{\partial q_x}{\partial x} + \frac{\partial q_y}{\partial y} = 0 \tag{8.1}$$

X-momentum conservation:

$$\frac{\partial q_x}{\partial t} + \frac{\partial (uq_x)}{\partial x} + \frac{\partial (vq_x)}{\partial y} + gh\frac{\partial (h+z)}{\partial x} + \frac{gn^2\|\vec{q}\|q_x}{h^{7/3}} = 0 \tag{8.2}$$

Y-momentum conservation:

$$\frac{\partial q_y}{\partial t} + \frac{\partial (vq_y)}{\partial y} + \frac{\partial (uq_y)}{\partial x} + gh\frac{\partial (h+z)}{\partial y} + \frac{gn^2\|\vec{q}\|q_y}{h^{7/3}} = 0 \tag{8.3}$$

This version of the equations is simplified by assuming the convective component of acceleration (second and third terms in equations 8.2 and 8.3) is negligibly small, which has been demonstrated to be a reasonable assumption for shallow, gradually varied flows where the Froude number remains below 0.5 (de Almeida and Bates 2013). The simplified 2D local inertial equations are (de Almeida and Bates 2013):

Mass conservation:

$$\frac{\partial h}{\partial t} + \frac{\partial q_x}{\partial x} + \frac{\partial q_y}{\partial y} = 0 \tag{8.4}$$

X-momentum conservation:

$$\frac{\partial q_x}{\partial t} + gh\frac{\partial (h+z)}{\partial x} + \frac{gn^2\|\vec{q}\|q_x}{h^{7/3}} = 0 \tag{8.5}$$

Y-momentum conservation:

$$\frac{\partial q_y}{\partial t} + gh\frac{\partial (h+z)}{\partial y} + \frac{gn^2\|\vec{q}\|q_y}{h^{7/3}} = 0 \tag{8.6}$$

As can be seen in these equations, convective acceleration terms in the x and y Cartesian directions [$\partial(uq_x)/\partial x$, $\partial(vq_x)/\partial y$, $\partial(uq_y)/\partial x$ and $\partial(vq_v)/\partial y$] are assumed negligible, such that only local acceleration [$\partial q_x/\partial t$ and $\partial q_y/\partial t$] is considered. Similarly, molecular and turbulent momentum diffusion are also neglected.

From equations 8.1 to 8.6 , $\vec{q}$ = unitary flow [m^3/(s-m)], with components q_x and q_y, h = flow depth, z = bed level, g = gravity, u = depth-averaged velocity in the x Cartesian direction, v = depth-averaged velocity in the y Cartesian direction, and n = Manning's roughness coefficient.

The explicit, staggered-grid, finite difference versions of these equations are (de Almeida and Bates 2013):

$$\eta_{i,j}^{n+1} = \eta_{i,j}^{n} + \frac{\Delta t}{\Delta x^2}\left(q_{i-1/2,j}^{n+1} - q_{i+1/2,j}^{n+1} + q_{i,j-1/2}^{n+1} - q_{i,j+1/2}^{n+1}\right) \tag{8.7}$$

$$q_{i-1/2,j}^{n+1} = \frac{\theta q_{i-1/2,j}^{n} + \frac{1-\theta}{2}\left(q_{i-3/2,j}^{n} + q_{i+1/2,j}^{n}\right) - gh_f^n \frac{\Delta t}{\Delta x}\left(\eta_i^n - \eta_{i-1}^n\right)}{1 + gn^2 \Delta t \|\vec{q}_{i,j}^{n}\|/h_f^{7/3}} \tag{8.8}$$

$$q_{i,j-1/2}^{n+1} = \frac{\theta q_{i,j-1/2}^{n} + \frac{1-\theta}{2}\left(q_{i,j-3/2}^{n} + q_{i,j+1/2}^{n}\right) - gh_f^n \frac{\Delta t}{\Delta x}\left(\eta_i^n - \eta_{i-1}^n\right)}{1 + gn^2 \Delta t \|\vec{q}_{i,j}^{n}\|/h_f^{7/3}} \tag{8.9}$$

These are the 2D mass and momentum conservation equations used by LISFLOOD-ACC, which are solved on a regular Cartesian grid ($\Delta x = \Delta y$).

Note that no momentum exchange takes place between the x and y directions (since convective acceleration is neglected), these being coupled solely through the mass conservation equation (equation 8.7).

Also, no lateral friction is considered, which is a consequence of the fact that water can move only in the two Cartesian directions. Water depths are evaluated at cell centers (i, j), whereas flows between cells are evaluated at cell interfaces ($i \pm 1/2$, $i \pm 3/2$, $j \pm 1/2$, $j \pm 3/2$).

In equations 8.7 to 8.9,

- $\eta = h + z$ is the water surface elevation, which results from adding bed elevation (h) and water depth (z)
- g is gravity
- h_f is the flow depth between two adjacent cells, defined as $max(\eta_i^n, \eta_{i-1}^n) - max(z_i, z_{i-1})$ (similarly for the y direction)
- q_x and q_y are the x and y components of the unitary flow vector $\vec{q}$ [m^3/(s-m)]
- Δt and Δx denote the temporal and spatial grid spacing (squared grid cells are assumed)
- n is Manning's roughness coefficient
- subscripts i and j refer to position in the x and y Cartesian directions, respectively. Fractional subscripts (e.g., $i + 1/2$, $i + 3/2$) refer to cell interfaces, whereas whole subscripts (e.g., $j, i + 1$) refer to cell centers
- superscript n refers to time step
- θ is a weighting factor that regulates the amount of artificial numerical diffusion (see de Almeida et al. 2012 for a detailed justification). This parameter was equal to 1 in all simulations performed here, so no artificial numerical diffusion was used.

The norm of the unitary flow vector in the denominator of equations 8.8 and 8.9, $\|\vec{q}_{i,j}^{n}\|$, is calculated as (de Almeida and Bates 2013)

$$\|\vec{q}_{i,j}^{n}\| = \sqrt{\left(q_{x,i,j+1/2}^{n}\right)^2 + \left(q_{y,i+1/2,j}^{n}\right)^2} \tag{8.10}$$

where

$$q_{x,i,j+1/2}^{n} = \frac{1}{4}\left(q_{x,i-1/2,j}^{n} + q_{x,i+1/2,j}^{n} + q_{x,i-1/2,j+1}^{n} + q_{x,i+1/2,j+1}^{n}\right) \tag{8.11}$$

and

$$q_{y,i+1/2,j}^{n} = \frac{1}{4}\left(q_{y,i,j-1/2}^{n} + q_{y,i,j+1/2}^{n} + q_{y,i+1,j-1/2}^{n} + q_{y,i+1,j+1/2}^{n}\right) \tag{8.12}$$

Here, $q_{xi,j+1/2}^{n}$ and $q_{y,i+1/2,j}^{n}$ are the x and y components of vector $\|\vec{q}_{i,j}^{n}\|$ at time step n, respectively. These components are used in the calculation of the friction term of the momentum equations, so they are located at the upper $(i, j+1/2)$ and right-hand face $(i+1/2, j)$ of cell i,j.

This numerical scheme is computationally more stable than the original formulation of Bates et al. (2010), and has been shown to be applicable for shallow, gradually varied flows where inertial forces do not play a predominant role ($Fr<0.5$, de Almeida and Bates 2013, de Almeida et al. 2012).

A compiled version of LISFLOOD-ACC was kindly provided for this dissertation by the School of Geographical Sciences of the University of Bristol.

The use of reduced complexity models in river science is discussed extensively in fields such as fluvial geomorphology (e.g., Brasington and Richards 2007), flood risk management (e.g., Bates et al. 2010) and riverine landscape ecology (e.g., Poole 2002).

Interestingly, although these have traditionally been separate bodies of literature, the need for computationally efficient and accurate hydrodynamic simulation has led them to converge regarding the use of reduced complexity models at the 'intermediate' scales required for river management (Brasington and Richards 2007). A review of this issue is presented by Coulthard and Van De Wiel (2012).

The reach-scale simulations performed here with LISFLOOD-ACC fall within the range of application of reduced complexity models discussed in the aforementioned fields. Further, within-channel habitat simulation with this family of models has been advocated, although only occasionally, in the cellular modeling literature (Nicholas 2009).

It has also been shown that reduced complexity models compare well with hydrodynamic numerical models of higher complexity at this scale, particularly regarding variables such as flow depth, unit discharge and energy losses (Nicholas 2009, Thomas and Nicholas 2002).

Thus, the computational overhead resulting from the use of fully dynamic hydraulic models for long-term reach-scale simulations can be avoided, which is fundamental to the applicability of the approach proposed in this thesis.

8.2.4.5 Model calibration

Models were calibrated by comparing water depths predicted by LISFLOOD-ACC and the field measurements of Januschke et al. (2014), which were distributed in cross sections approximately 20 m apart covering the entire study reach.

Calibration was done running steady state simulations with a warm-up period of one hour. The upper boundary condition for these simulations was the discharge value of the corresponding sampling day in the surveys of Januschke et al. (2014). Figure 8.7 shows the location of the calibration points in LISFLOOD-ACC's computational grid.

In the calibration process, Manning's roughness coefficient was varied until the best possible agreement between predicted and observed water depths was achieved.

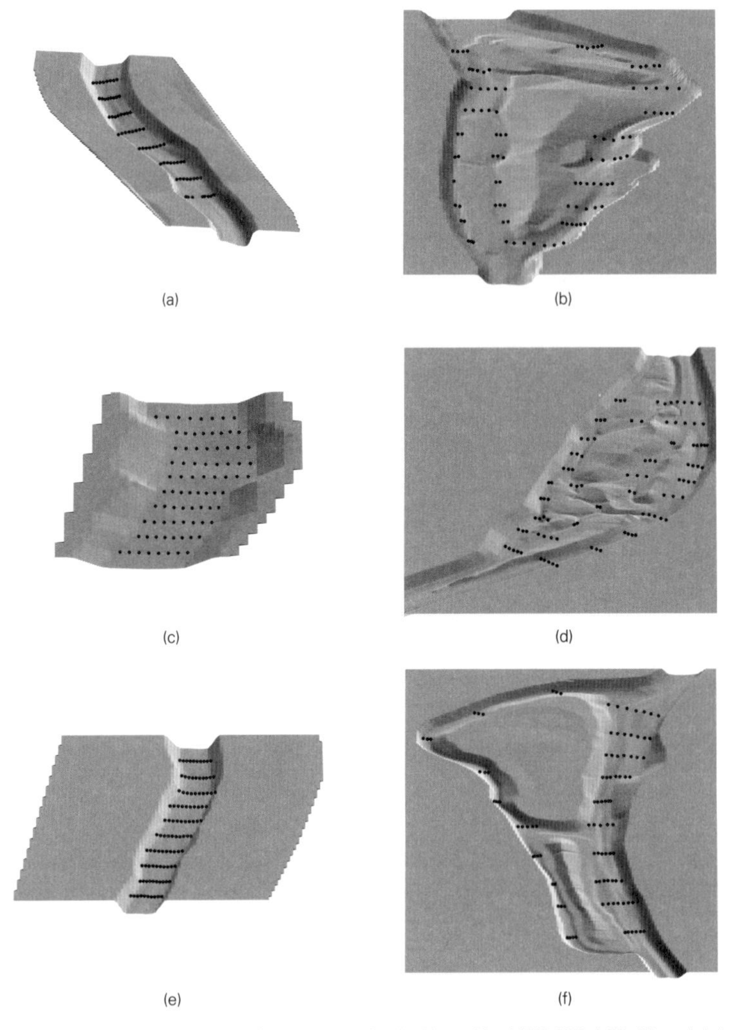

Figure 8.7: Three-dimensional view of the computational grid used by LISFLOOD-ACC. All models have been rotated and the vertical scale exaggerated to improve visualization. Calibration points are shown as black dots. These correspond to the points of the field survey conducted by Januschke et al. (2014). (a) Cölbe trained, (b) Cölbe restored, (c) Ludwigshütte trained, (d) Ludwigshütte restored, (e) Wallau trained, (f) Wallau restored.

This was assessed visually by plotting predicted water depths in the x axis and observed water depths in the y axis, and formally by testing ($\alpha = 0.05$) the null hypothesis that the slope of the fitted least-squares regression line is one and its intercept is zero (Piñeiro et al. 2008). The mean and standard deviation of the distances from each point to the 1:1 line were used to assess the size of prediction errors.

The resulting Manning coefficients are shown in figure 8.6. For all sites, a satisfactory agreement between measured and estimated water depths was achieved through increases in n between 50% and 100% relative to the starting values.

For channel areas, the resulting n values were between 0.035 (site ct) and 0.052 (site lt). Values between 0.095 and 0.1 were used for vegetated areas (grasses, shrubs and trees) and floodplains in restored sites, whereas the adjacent agricultural floodplain in trained sites was allocated values between 0.040 and 0.060.

Most of the calibrated Manning coefficients correspond well with those reported by Te Chow (1959) for small (width < 100 ft) natural upland streams with no in-channel vegetation and gravel/-cobble beds (n between 0.030 and 0.050, channel type D-1.b.1, page 111). However, some may be somewhat higher (e.g., sites lt and wr), which may be related to the fact that the local inertial approximation used by LISFLOOD-ACC may tend to slightly overestimate flow velocities when Fr increases to values close to 0.5 (Bates, P., pers. comm).

Despite this, results are in good general agreement with both observations and literature, and can be considered adequate for the habitat simulations undertaken here.

The resulting best-fit agreement between observed and predicted depth values is shown in figure 8.8. On visual inspection, the models show a good agreement with observed data, with 89% of the residuals lying within ± 10 cm for ct, 37% for cr (92% within ± 20 cm), 98.7% for lt, 92% for lr, 73% for wt (87% within ± 15 cm), and 83% for wr.

The confidence intervals for the slope and intercept (shown in figure 8.8) contained one (for the slope) and zero (for the intercept) in the cases of ct and lt, although all other models showed only slight deviations from consistency (*sensu* Piñeiro et al. 2008) (minimum consistency was 0.745 ± 0.131, the estimated slope for wr) and a small bias (maximum bias was 0.129 ± 0.037, the estimated intercept for cr).

These results can be considered satisfactory considering that a set of points scattered throughout the modeled reach make a more stringent test on the models than a comparison of water level profiles along the channel center line, as is commonly done in river modeling practice.

Further, the distribution of residuals mentioned above indicates that, for most areas of the studied reaches (e.g., shores, mid-channel, pools, shallow banks), the models are able to estimate water depths to within 10 - 15 cm.

For a spatial discretization using regular 1 m x 1 m cells this can be considered appropriate, although for shoreline habitats this may lead to inaccuracies and more precise alternatives should be explored (e.g., considering the depth in shore areas as subgrid phenomenon).

Unfortunately, as is commonly the case, no calibration data were available for high discharges, such that it must be assumed that the model performs equally well under these conditions. Uncertainty at higher flows can arise as the result of new areas being flooded which were dry for the calibration discharge, an increased role of turbulence (e.g., flow through vegetation) and secondary flows.

However, these effects are expected to be captured by the model, at least in part, through the use of literature-derived values of the roughness coefficient (*n*) (Te Chow 1959, USACE 1993) for overbank areas and islands, as well as the fact that LISFLOOD-ACC has been developed and rigorously benchmarked for high-flow conditions (de Almeida and Bates 2013, Neal et al. 2012).

Similarly, reduced-complexity models have been shown to perform well in terms of the generation of realistic two-dimensional flow patterns over beds with complex topography (Thomas and Nicholas 2002), which corresponds exactly with the purpose of the numerical simulations used in this methodological framework.

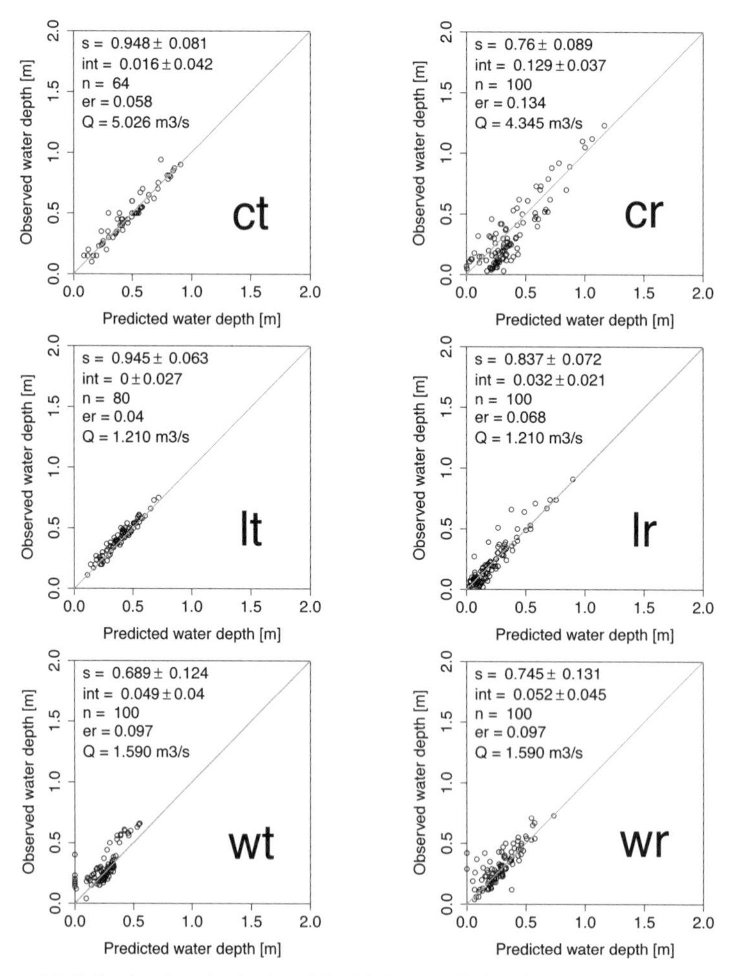

Figure 8.8: Calibration plots showing the relationship between the best-fit predicted and observed water depths for all models. s = estimated slope ± 2.53% confidence interval, int = estimated intercept ± 2.53% confidence interval, n = number of observations, er = root-mean-square error with respect to the 1:1 line, Q = discharge at time of calibration (m^3/s); cr = Cölbe restored, ct = Cölbe trained, lr = Ludwigshütte restored, lt = Ludwigshütte trained, wr = Wallau restored, wt = Wallau trained.

Borrowing from discussions in modern fluvial geomorphology (Bokulich 2013, Brasington and Richards 2007, Coulthard and Van De Wiel 2012), the aim here is not to model the exact trajectory of the system but rather to capture the essential properties of its long-term reach-scale dynamics, the purpose being to provide a description of its hydromorphological template (Poff and Ward 1990) that can then be associated with the survival chances of the populations of interest.

As discussed in sections 6.2 and 6.3, this description is based on the statistical properties of this dynamics, which are less sensitive to local inaccuracies in modeled water depths and/or flow velocities.

Further, higher accuracy levels in hydraulic simulation may not lead to more precise ecological analyses at this scale (see discussion in section 6.2.2). For these, as in fluvial geomorphology (Coulthard and Van De Wiel 2012), nonlinearity and stochasticity mediate relationships between state variables (Pickett et al. 2007), making deterministic approaches much less practicable than in other areas of river science such as hydraulic engineering and infrastructure design. Modeling strategies advocated in modern riverine landscape ecology (e.g., Poole 2000, 2002) reflect this perception.

8.2.5 Designating habitable space

Velocities and water depths from the hydraulic simulations were combined with the punctual substrate information collected by Januschke et al. (2014) in order to provide an instantaneous (daily) suitability value at each point.

The points used in this analysis were the same as in model calibration, which also correspond to the points where substrate (and water depth) data were collected by Januschke et al. (2014) (see figure 8.7).

As discussed in the introduction to this chapter (section 8.1), habitable space refers here to aquatic space in the study reach whose mechanical properties, resulting from local topography, substrate, flow velocity and water depth, make it suitable for habitation at a specific point in time and space.

Spatially, these points are defined at the same resolution as the hydraulic model. In this case, a point is assumed to represent the characteristics of a 1 m by 1 m grid cell (see section 8.2.4).

Habitale space was defined using macroinvertebrate shear stress tolerance values presented in Wolter et al. (2013). These authors reviewed a large number of laboratory and field studies in order to provide a compiled data base of the 'preferred shear stress environment' of a large number of taxa.

Shear stress was used given that it allows accounting for the interaction between flow velocities and substrate near the bottom, and also because the FST (FliesswasserStammTisch) hemisphere technique (Statzner and Müller 1989) provides a standardized methodology that improves comparability (Wolter et al. 2013).

Table 8.5 summarizes the 'preferred shear stress environment' and substrate associations of each species used in the analysis. Substrate associations were consulted at http://www.freshwater-ecology.info (Schmidt-Kloiber and Hering 2012) and specialized literature (Peeters et al. 2002, Schröder et al. 2013). For a detailed description of substrate types, see Hering et al. (2003).

Flow velocities and water depths computed by LISFLOOD-ACC (see section 8.2.4) were used as input for Bezzola's method (see section 7.2.2), which outputs an adjusted bottom shear stress based on substrate characteristics and depth-averaged flow velocity.

Habitable space was defined as a binary variable (i.e., 0 = unsuitable, 1 = suitable). No intermediate values were considered in attention to the levels of uncertainty typically found in habitat association models derived from observational studies (Lancaster and Downes 2010a,b).

The rule used to define habitability was the following:

> if the bezzola-adjusted bottom shear stress did not exceed the required value AND any of the required substrate types was present at a point, the point was considered suitable (habitat suitability = 1), otherwise, it was considered unsuitable (habitat suitability = 0).

A habitat simulation was run for each species at each site (cr = Cölbe restored, ct = Cölbe trained, lr = Ludwigshütte restored, lt = Ludwigshütte trained, wr = Wallau restored, wt = Wallsu trained), which resulted in a total of 15 species x 6 sites x 100 points per site = 9000 time series of binary habitat suitability. These were used to derive the habitat dynamics indicators proposed in the method (section 8.2.6).

8.2.6 Computation of habitat dynamics indicators using punctual substrate data - hsAv

In order to test the research hypothesis described in section 8.2.2, an indicator of the proportion of reach space that was habitable during the study period and its temporal variability was used.

Table 8.5: 'Preferred shear stress environment' and substrate associations of the species used in the habitat suitability simuations. Substrate associations taken from Schmidt-Kloiber and Hering (2012) and Schröder et al. (2013). Shear stress requirements from Wolter et al. (2013). Abbreviations of substrate types: subM = submerged macrophytes, LPTP = living parts of terrestrial plants, CPOM = coarse particulate organic matter, FPOM = fine particulate organic matter.

Species	**Preferred shear stress [N/m^2]**	**Substrate associations**
Allogamus auricollis	0.19	Psammal, Akal, Mesolithal, Mikrolithal, Makro-Megalithal, subM, LPTP, CPOM, FPOM
Anabolia nervosa	0.20	subM, LPTP, CPOM, FPOM, Xylal
Asellus aquaticus	0.23	Pelal, Argyllal, Psammal, Akal, Mesolithal, Mikrolithal, Makro-Megalithal, subM, LPTP, CPOM, FPOM
Athripsodes albifrons	0.44	Akal, Mesolithal, Mikrolithal
Athripsodes cinereus	0.54	Psammal, Akal, Mesolithal, Mikrolithal
Baetis fuscatus	0.52	Psammal, Akal, Mesolithal, Mikrolithal, Makro-Megalithal, subM, LPTP, CPOM, FPOM
Baetis lutheri	1.35	Akal, Mesolithal, Mikrolithal, Makro-Megalithal, subM, LPTP
Caenis luctuosa	0.21	Pelal, Argyllal, Psammal, Akal, Mesolithal, Mikrolithal, Makro-Megalithal, subM, LPTP, CPOM, FPOM
Ecdyonurus venosus	0.46	Akal, Mesolithal, Mikrolithal, Makro-Megalithal, subM, LPTP, CPOM, FPOM
Ephemera danica	0.26	Pelal, Argyllal, Psammal, Akal, Mesolithal, Mikrolithal, Makro-Megalithal, subM, LPTP
Goera pilosa	0.34	Mesolithal, Mikrolithal, Makro-Megalithal
Habrophlebia lauta	0.27	Pelal, Argyllal, Psammal, Akal, Mesolithal, Mikrolithal, Makro-Megalithal, subM, LPTP, CPOM, FPOM
Odontocerum albicorne	0.43	Psammal, Akal, Mesolithal, Mikrolithal, Makro-Megalithal, subM, LPTP, CPOM, FPOM, Xylal
Potamanthus luteus	0.62	Pelal, Argyllal, Psammal, Akal, Mesolithal, Mikrolithal, Makro-Megalithal, subM, LPTP, CPOM, FPOM
Psychomyia pusilla	0.64	Mesolithal, Mikrolithal, Makro-Megalithal

This indicator is denoted here as hsAv (see section 6.2.4.3).

First, the fraction of habitable points was estimated at all time steps, which was simply the sum of the suitability of all points divided by the number of points (100). Then, an empirical cumulative distribution function (ecfd) of this new time series was built using R's function 'ecdf'. Relative frequency was used in order to guarantee comparability between species, sites and simulation lengths.

Reaches with a low availability of habitable space will exhibit an ecdf that rapidly grows to one, i.e., passes close to the upper left corner of the diagram. This results from time steps with a low fraction of habitable space having the highest relative frequencies.

In contrast, reaches with a high availability of habitable space will exhibit an ecdf that grows more slowly and thus passes closer to the lower right corner of the diagram. This results from time steps with larger fractions of habitable space having the highest relative frequencies.

Finally the area above the ecdf was calculated using R's package 'sp', which provides a number ranging between zero, when no habitable points were available during the study period, and one, which means that all points were habitable all of the time.

This number is denoted here as hsAv, and it is the habitat dynamics indicator underlying the analyses presented below. For a graphical illustration of this indicator, see figure 8.9.

The area above the ecdf is related to the temporal availability of habitable space, which is understood here in a manner analogous to the computation of the Gini coefficient in economics (Gini 1997). A small area above the ecdf curve (value closer to zero) suggests that most of the time (days of the year) only a small fraction of the instream area is habitable, whereas a large area (value closer to one) suggests the opposite.

Table 8.6: Kruskall-Wallis chi-squared test results for the differences between bottom shear stress in the trained and restored reaches. All tests were conducted with 1 degrees of freedom.

Site	Kruskall-Wallis χ^2	p-value
Cölbe	55958.95	2.20e-16
Ludwigshütte	11372.06	2.20e-16
Wallau	195797.3	2.20e-16

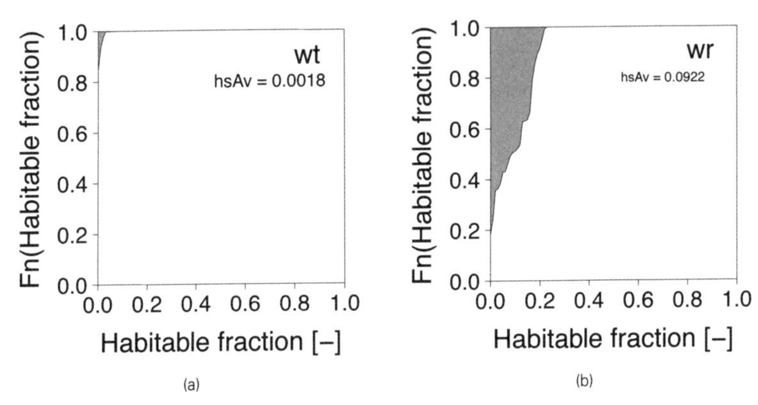

Figure 8.9: Empirical cumulative distribution function (ecdf) for the fraction of points habitable by *Goera pilosa* in sites wt (a) and wr (b). The different areas above the curves in wt and wr indicate the difference between the restored and trained situation.

8.3 RESULTS

8.3.1 Dynamics of aquatic habitable space in trained and restored reaches

Overall, the amount of habitable space (habitable fraction of points) was higher in restored than in trained reaches. An example is shown in figures 8.11 (a) and (b), where it can be seen that, on the same date, the habitable fraction of reach area for *Odontocerum albicorne* is higher in site wr than in wt.

This tendency was maintained throughout the study period, as in shown by the time series of habitable fraction shown in figures 8.11 (c) and (d). Similar results were found for the other four sites (ct, cr, lt, lr).

Pooling together results for all (15) species, restored and trained reaches exhibit significantly different values of hsAv (Kruskal-Wallis chi-squared = 60.08, df = 1, p-value = 9.121e-15), with restored reaches having a higher habitable space availability than trained ones.

A boxplot of these results is shown in figure 8.10. Despite being significantly higher than in trained reaches, hsAv in restored sites was always below 0.26 (i.e., 26% of the 100 points sampled).

These differences are directly attributable to the geometrical (and therefore also hydrodynamic) and substrate differences created by the restoration projects the in restored sites. Significantly more substrate types (Kruskal-Wallis chi-squared = 3.97, df = 1, p-value = 0.046) and higher substrate diversity (Shannon's H, Kruskal-Wallis chi-squared = 3.86, df = 1, p-value = 0.049) were found in restored relative to trained sites (see figure 8.12).

Similarly, the distribution of simulated, Bezzola-adjusted bottom shear stress (figure 8.13) shows that trained sites had, in general, higher values than restored ones, although occasional localized peaks were observed in both (up to 290 N/m^2 in cr).

These differences between trained and restored sites were confirmed statistically (table 8.6), although the high significance values are clearly related to the large number of observations in the tests.

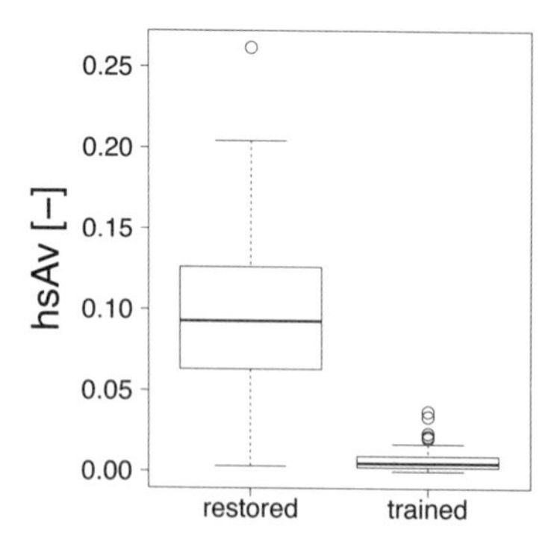

Figure 8.10: Boxplot comparing the values of hsAv for all species in trained and restored reaches. Differences are statistically significant (Kruskal-Wallis chi-squared = 60.08, df = 1, p-value = 9.12e-15).

8.3.2 Relationship between habitable space dynamics and macrozoobenthos community

At first glance, a bivariate plot of the relationship between hsAv and average density [ind./m^2] [figure 8.14(a)] shows no apparent trend over the full range of hsAv. However, closer inspection reveals clustering of points towards the lower left corner of the plot [figure 8.14(b)].

This tendency indicates that there is a concentration of taxa whose average local density is low at the lower end of the habitat availability (hsAv) range found in the study area.

If the data are re-plotted for the hsAv interval shown in the grey box in figure 8.14(a) (0.000 - 0.021), the distribution of points seems to indicate that for lower values of hsAv, only low values of avgN occur, and as hsAv increases, higher local average densities start to appear. This can be

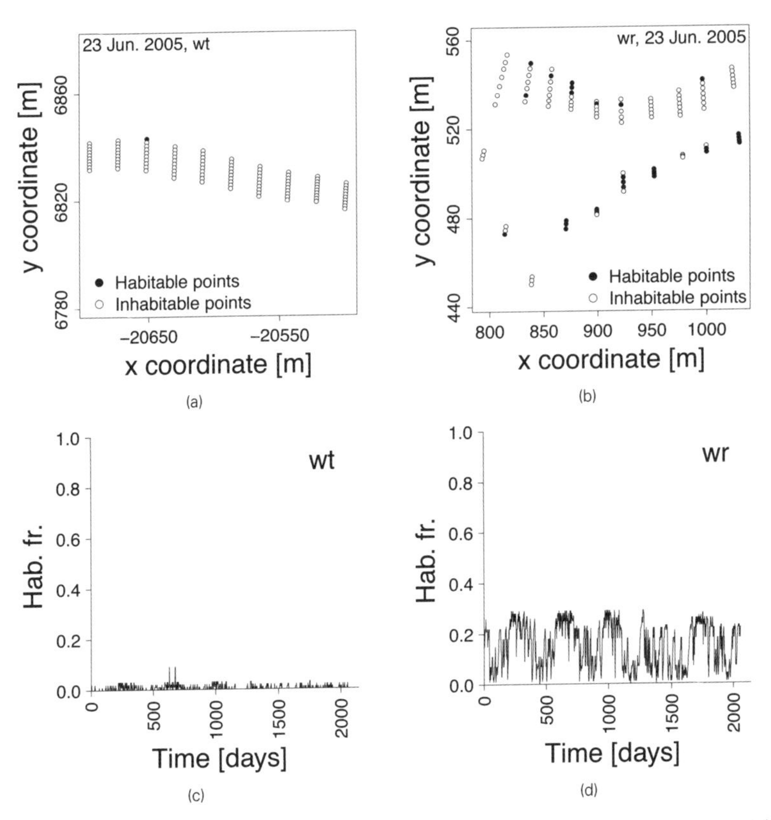

Figure 8.11: Habitable points for species *Odontocerum albicorne* in sites (a) Wallau trained (wt) and (b) Wallau restored (wr) on the same date (June 23, 2005). Habitable fraction in wt = 0.01, in wr = 0.29 . The aspect ratio of both plots has been distorted ($\neq 1$) to improve the visibility of points. x and y coordinates refer to a local coordinate system. (c) and (d) show the corresponding time series of habitable fraction of points ("Hab. fr." on the y-axis).

interpreted as suggesting a limiting-type relationship at this range of habitable space availability (hsAv).

This trend can be further explored graphically by introducing resVal as a covariate and plotting avgN vs. hsAv for different levels (intervals) of resVal. Such a diagram, referred to in the R jargon as a 'coplot' (Cleveland 1993), is shown in figure 8.15. This graph shows that, for increasing levels of resistance (resVal), different behaviors are exhibited by local average density as hsAv increases.

For the first two plots in figure 8.15 (from left to right), lower values of avgN are observed throughout the examined range of habitat availability (hsAv), except for two outliers (*Athripsodes albifrons* and *A. cinereus* in site ct, see discussion in section 8.4).

In the third graph, higher avgN values occur beyond hsAvs of 0.2% - 0.3%. With yet higher resVal values (plot on the right), such high (relative to the first two plots) average local densities (avgN) are also achieved, and a smaller relative proportion of points near the x axis towards the lower end of the hsAv range can be observed.

Similar trends were observed for different groupings of resVal, showing that these findings are not simply the result of the bins used in figure 8.15.

These results indicate that more resistant taxa tend to exhibit a steeper increase in their average local density than less resistant ones once a threshold of habitat availability (hsAv) has been reached. The presence of low average local densities throughout the plotted range of hsAv suggests, as expected, that responses to habitat availability are of limiting rather than central nature (*sensu* Lancaster and Belyea 2006).

Grouping observations by the aforementioned levels of resVal reveals a similar trend (figure 8.16). The second and third quartiles of the distribution within each resVal level also tend to increase with this variable, whereas the lower end of the distribution (e.g., the first quartile) remains close to the x axis throughout.

A formal statistical assessment of the above tendencies using quantile regression (Koenker 2015) (8.7) indicates the following. For the first resVal bin, neither the intercept nor the slope of

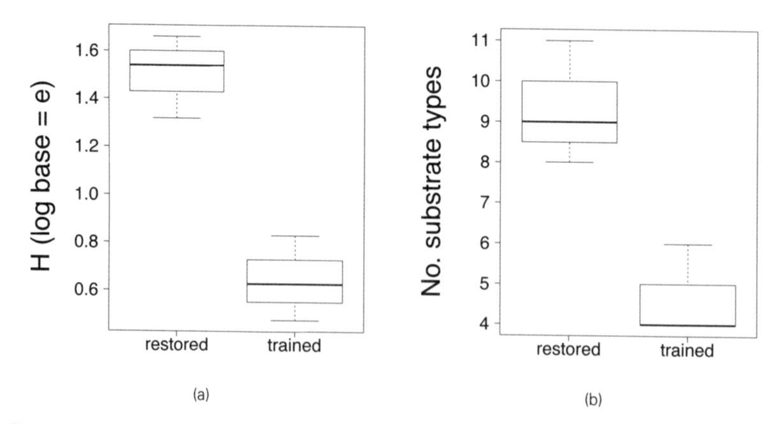

(a) (b)

Figure 8.12: Boxplots of (a) Shannon diversity (H, calculated with natural logarithm) for substrate and (b) number of substrate types in trained and restored sites.

the quantile regression for $\tau = 0.75$ are significantly different from zero. For the second bin ($0.49<\text{resVal}\leq0.56$), only the intercept is significantly greater than zero.

The third resVal group ($0.56\leq\text{resVal}<0.61$) exhibits no significant differences from zero, although the p-value for the slope is relatively low. p-values far from significance result again for the last resVal bin ($\text{resVal}\geq0.61$), with 0.946 for the intercept and 0.370 for the slope.

Effect sizes shown in table 8.7 suggest much higher slopes are possible for the third and fourth resVal bins despite the uncertainty (se). The estimated ranges for this quantity in all bins are very wide; however, the position of these intervals relative to one another increases with resVal.

This can be interpreted as an indication that less resistant species tend to be limited at lower values of hsAv, and that this limitation tends to break down as resistance (resVal) increases. Visual inspection of figure 8.15 indicates that, overall, higher local average densities (avgN) at low hsAv values are achieved only in the third and fourth resVal bins, i.e., by more resistant species.

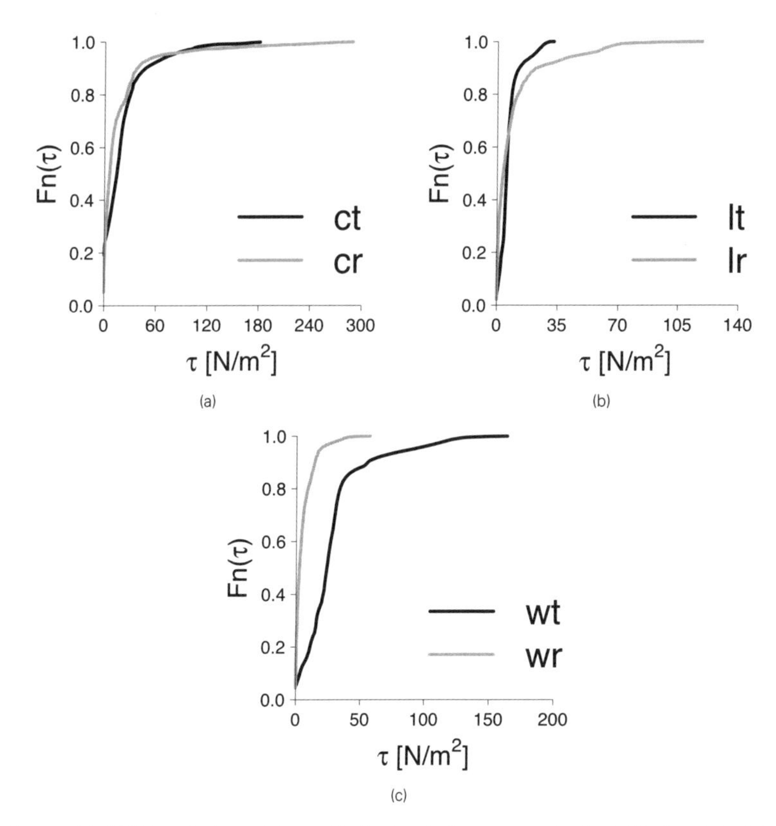

Figure 8.13: ECDF (empirical cumulative distribution function) curves showing the distribution of simulated, Bezzola-adjusted bottom shear stress in (a) Cölbe, (b) Ludwigshütte and (c) Wallau. Note that trained sites tend to exhibit higher values of this quantity than restored ones. Statistical differences were significant ($p<0.05$), see text. ct = Cölbe trained, cr = Cölbe restored, lt = Ludwigshütte trained, lr = Ludwigshütte restored, wt = Wallau trained, wr = Wallau restored.

A Kruskal-Wallis test for the data shown in figure 8.16 indicates differences that are close to statistical significance (Kruskal-Wallis chi-squared = 5.3758, df = 3, p-value = 0.1463). This result also suggests ecologically meaningful differences between the response of species with different degrees of resistance to habitable space scarcity.

8.4 DISCUSSION

The fact that hsAv be significantly higher in restored (median = 0.0929) than in trained (median = 0.00501) sites is an indication of the ability of the method to discriminate between these two hydromorphologically contrasting conditions. The magnitude of this effect is not trivial, as is shown by the results of the Kruskall-Wallis test and the difference between the median hsAv in trained and restored sites.

The characteristics of the point distribution observed in figure 8.15 and the results of the quantile regressions suggest the existence of a limit imposed on local average density (avgN) by habitable space availability (hsAv).

The exact position of this limit in this study, however, is hard to ascertain given the uncertainty associated with the preferred shear stress values used in the habitat simulations. These values stem, to an important degree, from observational field studies (*sensu* Quinn and Keough 2002), and it is risky to set an upper limit on the mechanical tolerance of macroinvertebrates based on measurements made at a single discharge (i.e., that of the measurement day).

Moreover, the capacity of these organisms to cope with mechanical stress is the result of a wide variety of morphological, physiological and behavioral strategies (Lancaster and Downes 2013).

Hence, such limits may oversimplify the actual tolerance of the target taxa. If that were the case, then the hsAv threshold observed in this study may underestimate the actual capacity of the studied species to cope with hydrodynamic stress. Further, this may help explain why the limitation was observed at such low values of hsAv (0.2% - 0.3%).

Another point that warrants further examination is the observation that, at the scale of the studied reaches (width between 30 - 50 m, length between 150 and 200 m, model elements 1 m x 1 m

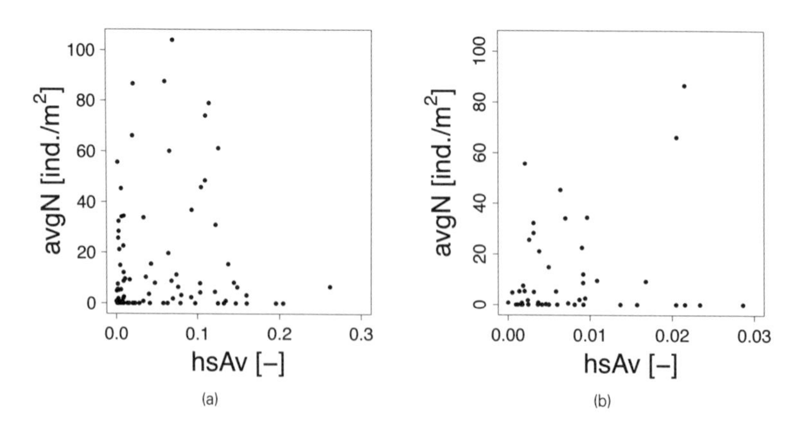

Figure 8.14: (a) Local average density (avgN) vs. habitable space availability (hsAv) for the full range of the latter. (b) The x axis has been expanded by plotting only the range of hsAv values between 0.000 and 0.03.

in size) the local inertial approximation may tend to systematically, though only slightly, underestimate water depths (Bates, P., de Almeida, G., pers. comm) (figure 8.17). This may also contribute to hsAv being underestimated and therefore to a lower position of the estimated threshold value.

Despite these limitations, the purpose of the test application was fulfilled in regards to the existence of the aforementioned limit, as suggested by the graphical exploration of figure 8.15 and the results of the quantile regressions.

The fact that, collectively, less resistant taxa (low resVal values) exhibit lower average local densities (hsAv) at low hsAv values, and, concurrently, more resistant species are able to build larger populations at the same hsAv levels, strongly suggests that the former are limited by the spatiotemporal availability of mechanically habitable space.

If this phenomenon were observed for just a single taxon, much less confidence could be placed in such a conclusion. However, the fact that many taxa collectively exhibit this behavior can be understood as strong evidence (Downes 2010) in favor of the research hypothesis presented in section 8.2.2.

Table 8.7: Quantile regression results (first resVal bin without outliers). se = standard error. All statistics were calculated with the bootstrap option of R's function rq (package quantreg, Koenker 2015).

resVal bin	Parameter	Value	se	Student's t	p-value
<0.49	Intercept	0.44	2.31	0.19	0.854
	Slope	216.67	344.40	0.629	0.547
0.49 - 0.56	Intercept	4.15	1.64	2.52	0.026
	Slope	305.13	345.17	0.88	0.394
0.56 - 0.61	Intercept	10.76	13.31	0.81	0.434
	Slope	2702.11	2429.35	1.11	0.286
>0.61	Intercept	-2.29	31.91	-0.07	0.946
	Slope	3804.39	3860.33	0.98	0.370

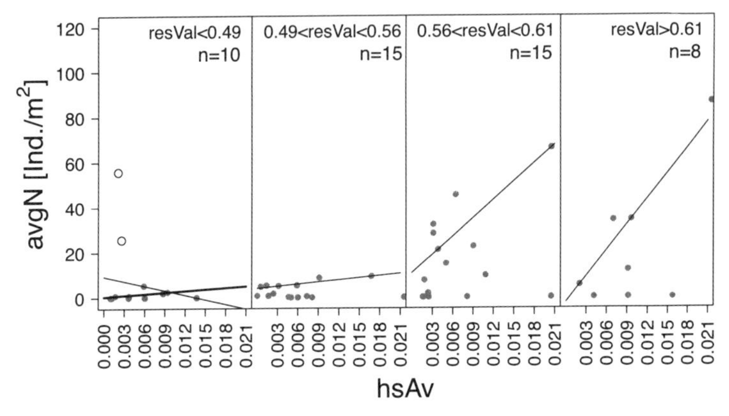

Figure 8.15: Coplot of averge local density (avgN) vs. habitable space availability (hsAv) given species resistance (resVal, abbreviated in the plots as rV), using only the lower portion of the x axis (from 0.000 to 0.0021). The line in each plot corresponds to a quantile regression (τ = 0.75) for the hsAv range 0.000 - 0.021. The two white points in the first (left) plot are outliers and the thicker line is the quantile regression without these points. n = number of points in each plot (without outliers for the first plot). See discussion for explanation.

The results of the quantile regressions were not statistically significant. However, effect sizes suggest ecologically meaningful differences in overall population success at increasing values of resVal along the observed gradient of hsAv.

A non-formal examination of the estimated regression parameters (table 8.7) indicates that, at low hsAv values, smaller populations were more common for the first two bins, as indicated by the low values of avgN at this resistance levels. The slope of neither of the τ = 0.75 quantile regression lines was greater than zero, and their p-values were far from significance (see table 8.7).

For the third resVal bin, a different situation was observed. The estimated slope, despite not being significantly different from zero, had a much greater value and a p-value closer to statistical significance (p-value = 0.286). This suggests that the achieved maximal local average densities tended to be higher, in comparison to less resistant taxa, as habitable space increased.

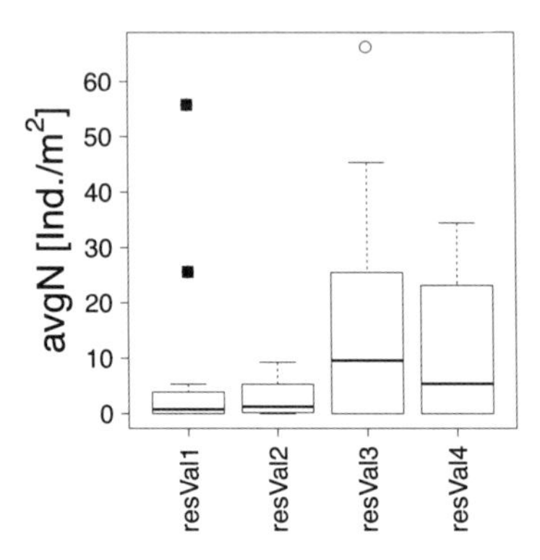

Figure 8.16: Boxplot of average local density (avgN) at different values of resVal. **resVal1** through 4 correspond to the bins used in figure 8.15 for this variable. Squares indicate the same outliers as in the leftmost plot of figure 8.15.

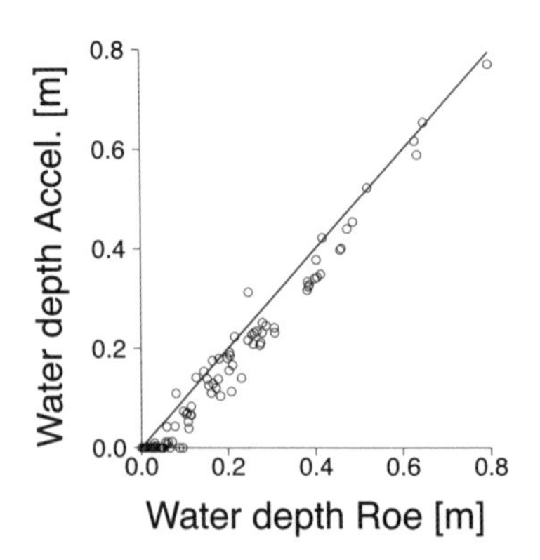

Figure 8.17: Comparison between water depths in site cr estimated with LISFLOOD using the Roe (full 2D equations) and acceleration ("Accel.", y-axis) schemes. Discharge is steady at Q = 0.909 m^3/s.

A similar situation was found for even higher values of species resistance (resVal), although the smaller n in this bin restricts conclusions somewhat. Nonetheless, the distribution of points in the plot does suggest a reduction of the limiting effect of habitat scarcity for species in this resVal bin as well.

These limiting-type relationships would be better represented at their lower end by nonlinear (perhaps logistic) curves. However, at this point, I opted for the simplest type of relationship (i.e., linear) given the uncertainties involved. More data would allow fitting more complex curves (e.g., logistic) with a higher degree of confidence.

The two outliers in figure 8.15 (white circles) correspond to the local average densities of *Athripsodes albifrons* and *A.cinereus* in site ct. Their inclusion in the quantile regression for this resVal bin does not change the trend already described, i.e., slope (-810.20 $\pm$ 3939.40, p-value = 0.842) and intercept (10.10 $\pm$ 26.73, p-value = 0.714) are not significantly different from zero.

However, outliers change the size of the effect in both sign and magnitude, from a very small positive slope to a steeper negative one, which is obviously caused by the higher avgN values of the outliers at low hsAvs. Again, it is possible that the preferred bottom shear stress values used in the habitat simulations (Wolter et al. 2013) tend to underestimate the actual tolerance to mechanical stress of these taxa, in which case these avgN values would in reality correspond to higher values of hsAv.

At the same time, the substrate information used does not account explicitly for the availability of flow refugia, which could allow populations to maintain higher local densities despite habitable space scarcity. Similarly, traits not accounted for in the computation of resVal could confer additional resistance to this taxa, which could help explain their observed success at low hsAv.

A nearby highly productive habitat patch not detected by the sampling scheme could also help explain this finding. The existence of outliers in ecological data is among the most important justifications for the use of quantile regression in this field (Lancaster and Belyea 2006), which, unlike central-type regressions, is less sensitive to the large leverage of outliers (Koenker and Hallock 2000).

In conclusion, it was possible to observe the limitations imposed on population success by the spatiotemporal availability of mechanically suitable space instream. This finding constitutes a confirmation of the ecological concepts that underlie the proposed methodological framework, which were derived from decades of observational and experimental studies.

The physical habitat template concept (Poff and Ward 1990) and its associated ideas (Hart and Finelli 1999, Pringle et al. 1988, Southwood 1977, Statzner et al. 1988, Townsend 1989, Townsend and Hildrew 1976), combined with appropriate statistical techniques (Lancaster and Belyea 2006), can reveal the limiting effect of the dissapearance of habitable space on population success of the selected macroinvertebrate taxa. This limitation varies along a gradient of species resistance, with a stronger limitation on less tolerant taxa.

Detailed laboratory analyses could provide more accurate estimations of the tolerance ranges of these organisms to mechanical stress, which could then provide better input for the habitat simulations proposed here. Similarly, substrate-specific sampling could be used to carry out a more detailed check of the research hypothesis addressed in this test application.

9 CONCLUDING REMARKS

This dissertation presents an approach for analyzing stream hydromorphology from an ecological point of view. It is based on a combination of ecological concepts on habitat-biota relations and numerical simulation tools available in fluvial hydraulics, which are able to provide a much needed spatially and temporally explicit description of habitat conditions within the stream.

Although numerical hydrodynamic simulations constitute the core of the proposed method, equal emphasis has been placed on its ecological background, which is unavoidable given that the response variables involved in this research question are ecological in nature. Furthermore, it is the belief of the author that truly interdisciplinary work in ecohydraulics can only take place when professionals from both disciplines are exposed to the principles and methods of both fluvial ecology and hydraulics.

This course of action allowed developing a method that goes beyond existing approaches for describing hydromorphology. These provide a static description of the stream's structural features and their degree of human intervention, and do so at a maximum resolution of hundreds of meters (100 m at most).

However, at least for a biological target group such as macrozoobenthos, such a description is still significantly coarser than the perception range of individual organisms. At the same time, it does not take account of the role of in-stream spatial and temporal variability in the long-term persistence of stream organisms, a phenomenon that has been well known to stream ecologists for several decades.

Thus, this dissertation aims to contribute to bridging these two, as yet largely disconnected bodies of knowledge, which is particularly important in light of the fact that the methods used in each of them complement rather than exclude each other.

The indicators proposed in the method reveal how much ecologically-relevant information can be derived from numerical shallow water models, which in traditional habitat modeling is restricted to an estimation of habitat area at particular discharges. This, however, requires that the models be set up following ecological criteria, specifically with respect to scale (spatial and temporal extent and resolution) and the hydraulic processes that must be simulated explicitly.

In this respect, the results of the proof-of-concept suggest that, as can be deducted from hierarchy theory (O'Neill 1986), it is not always necessary to simulate milimeter-scale turbulent structures in order to study the long-term persistence of aquatic invertebrates in a stream. This conclusion is in line with modern, anti-reductionist perspectives in fluvial geomorphology, where the use of reduced-complexity models as an epistemological tool was pioneered in fluvial science in the 1990s (Bokulich 2013).

The proof-of-concept also allowed successfully testing the rationale behind the approach as a whole, which provides a biological basis for the set of proposed hydromorphological indicators. At the same time, this successful test constitutes further evidence in support of the ecological concepts underlying the method.

In the context of WFD monitoring, the proposed analysis strategy can be implemented for a group of key type-specific species, so as to provide a species-specific measure of adequacy of the hydromorphological template. This approach would allow checking hydromorphological suitability for the set of species with the strongest impact on ecological status or potential.

From this, conclusions such as 'x% of WFD-relevant macrozoobenthos taxa have an adequate hydromorphological template' would be possible. However, care must be taken not to interpret this result as suggesting that habitat must be provided everywhere for all taxa at all times, since this would lead restoration and assessment efforts to wrongly favor 'stasis' (*sensu* Palmer et al. 1997).

Finally, it is important to mention that habitable space can disappear as a result of many different types of changes in hydromorphology: drying ($h \rightarrow 0$), mechanical effects of velocity and turbulence, and substrate changes (erosion/sedimentation, clogging of interstices), all entailing negative biological responses which can nonetheless be very different.

Hence, the indicators presented in this thesis must be understood as an alternative that aims to integrate (*sensu* Orians 1980) the effects of the above individual processes, at the scale of a stream reach and for all the individuals it may potentially host. The mechanistic details of each individual process remain unknown in this analysis, but, as shown in the proof-of-concept, the limitation pattern that emerges at this scale is consistent with the accumulated biological knowledge on the studied macrozoobenthos species.

10 OUTLOOK

This thesis provides the basis for future investigations in which the habitable space gradient is studied in further detail, as well as for a larger set of target species and stream types. Study designs involving substrate-specific macrozoobenthos samples, continuous substrate maps (e.g., raster) and calibration data for a wide spectrum of discharges will allow determining species- and type-specific thresholds for the proposed indicators.

Habitable space simulations can be extended to include sediment transport and morphological change. However, long-term data for calibration and validation must be available for this.

The results of the proof-of-concept suggest that tolerance ranges for many species are still poorly known or are based on anecdotal evidence. Hence, flume experiments are required in which this is studied in more detail.

Given the mismatch between the scales of perception of macrozoobenthos and the scales at which their long-term persistence must be studied, flume experiments for exploring the relationship between in-stream, spatially-averaged (tens of squared centimeters) flow conditions and those prevailing at the mm-cm-scale should be conducted. This would make it possible to link, at least qualitatively, flow patterns at the scale of individual organisms based on those normally produced by habitat models.

The informatic quality and performance (computation time) of the patch mapping, patch building, patch tracking and patch juxtaposition algorithms can be further improved, perhaps by porting the code to faster, non-interpreted programming languages (e.g., C++, FORTRAN).

The stream patch dynamics framework proposed in this dissertation can serve as a basis for examining species-specific survivorship thresholds using stochastic spatial population dynamics models.

Finally, the methodological framework proposed here for mechanically suitable space can be expanded to other factor groups affecting macrozoobenthos. Numerical water quality modeling, for instance, could allow simulation of oxygen and organic matter dynamics as a new layer of information on top of the structural and mechanical variables employed here.

11 R SCRIPTS

The R scripts in which the current version of this algorithm is written make use of the following R packages: akima (Akima et al. 2013), raster (Hijmans 2013), sp (Pebesma and Bivand 2005), maptools (Bivand and Lewin-Koh 2013), rgeos (Bivand and Rundel 2013), shapefiles (Stabler 2013), animation (Xie 2013), lattice (Sarkar 2008), igraph (Csardi and Nepusz 2006) and rgdal (Bivand et al. 2013).

The data for running the scripts presented here must be organized in separate folders containing input, scripts and output files, as follows:

- ...\hydromStress \input
- ...\hydromStress \output
- ...\hydromStress \scripts

Here, 'hydromStress' is the root directory for all data, although it can be renamed to anything else by the analyst as long as the corresponding changes are made in script 'control.R'. Folder '...\hydromStress \input' must contain the following data:

- Output from the numerical hydraulic simulation
  - Text file with u-velocity component
  - Text file with v-velocity component
  - Text file with velocity magnitude (vMag)
  - Text file with flow direction data
  - Text file with water depth
- Mask for clipping rasters (polygon shape file). This mask defines the boundary of the study area
- Substrate raster (in ESRI Grid format)
- Suitable substrate polygon file (in ESRI polygon shapefile format)
- Bounding polygon for seeding points in 'journeyTrack.R' (in ESRI polygon shapefile format)

The output of the numerical hydraulic model must be stored as comma-separated values, as follows:

```
x,y,h0,h1,h2,h3,h4,h5,h6,h7,h8,h9,h10
30246.500,50381.500,1.975,0.323,1.835,0.724,1.322,0.784,0.823,0.011,0.897,1.182,1.058
30263.000,50393.000,1.352,0.108,1.910,1.184,0.834,0.178,1.503,0.593,0.843,0.408,1.876
30279.000,50404.500,0.651,0.471,0.201,1.430,1.568,1.775,1.832,1.703,0.793,0.181,1.705
30295.500,50416.000,1.308,1.586,0.029,1.590,1.715,1.629,0.786,1.516,1.055,1.517,0.227
```

```
30312.000,50427.000,0.723,1.765,0.142,1.106,1.694,1.177,0.628,0.853,1.222,0.302,1.589
30329.000,50438.000,1.861,1.366,0.932,1.301,1.942,1.069,1.072,0.723,1.794,0.755,1.992
30348.000,50443.500,0.736,1.733,0.887,0.871,1.445,0.184,1.755,1.157,0.848,0.479,1.775
30367.000,50449.000,1.632,1.958,0.567,0.323,1.622,1.307,0.635,0.038,0.944,0.178,1.061
```

This example file contains eleven time steps (from h0 to h10) of variable h (other column names are allowed as long as they comply with R syntax) at eight points (lines). The first (header) line shows the column titles, whose order must match the one presented here. The first two columns correspond to the x- and y-coordinates of each point (nodes in the numerical hydraulic model), and the remaining columns hold the data for all time steps in the simulation.

The order of execution of the scripts is as follows:

1. control.R - sets the environment and pathnames for all input and output files
2. patchMaps.R - interpolates and combines hydraulic model results and substrate maps for all time steps to produce patch maps. One patch map is stored in '...\hydromStress \output' for each simulation time step. The file format is ESRI polygon shape file. If a proj4 string is provided, a .prj file will be produced for each patch map
3. patchTrack.R - performs the patch tracking step and generates the areas and centroids matrices
4. ind_area_analysis.R - calculates patch size and area loss indicators
5. ind_patch_duration.R - calculates patch duration indicators
6. ind_journeyTrack.R - performs the patch juxtaposition analysis (calculation of *c*)

After execution, all output files are stored in '...\hydromStress \output'. These can be grouped as follows:

- Output from 'patchMaps.R'
  - Patch map vector files (ESRI polygon shapefiles)
- Output from 'patchTrack.R'
  - 'areasMatrix.txt'
  - 'xCentroidMatrix.txt'
  - 'yCentroidMatrix.txt'
- Output from 'ind_area_analysis.R'
  - 'A.txt' - total habitable area
  - 'N.txt' - number of patches
  - 'I_Delta.txt' - total fraction of area lost
  - 'liSpatial.txt' - spatial distribution of pr_li_le.5 (the probability of a patch experiencing a size loss of 50% or less)
  - 'lossList.txt' - list of fraction of area losses for all patches and all time steps
- Output from 'ind_patch_duration.R'
  - 'interSpellDurList.txt' - list of the duration of all inter-spell periods in patch existence
  - 'patchDuration.txt' - patch-specific duration statistics
  - 'patchDurationList.txt' - list of the starting point and duration of all duration spells of all patches throughout the analysis period
  - 'uiSpatial.txt' - spatial distribution of patch duration statistics
- Output from 'ind_journeyTrack.R'
  - 'c.txt' - time series of *c*

R scripts:

control.R

```
1 # set environment and load required packages
2 options(max.print=5.5E8,width=10000,signif=20) # set console
    options
```

```
3 library(akima)
4 library(raster)
5 library(sp)
6 library(maptools)
7 library(rgeos)
8 library(shapefiles)
9 library(animation)
10 library(lattice)
11 library(igraph)
12 library(rgdal)
13 library(ggplot2)
14
15 # full path of ascii files for V, h, substrate (ascii raster) and
     suitable substrate map (ESRI shapefile)
16 vData<-"...\\hydromStress\\input\\vMag.txt"
17 flowDirData<-"...\\hydromStress\\input\\fd.txt"
18 hData<-"...\\hydromStress\\input\\h.txt"
19 substrateRaster<-"...\\hydromStress\\input\\s.asc"
20 suitableSubstratePoly<-"...\\hydromStress\\input\\s1.shp"
21
22 # full path for output (patchmaps [ESRI shapefiles], matrices, time
      series) and input
23 outDir<-"...\\hydromStress\\output\\"
24 inDir<-"...\\hydromStress\\input\\"
25
26 # set options for animations
27 ani.options(interval=.35,convert="c:\\ImageMagick-6.8.8-Q8\\convert
     .exe",verbose=FALSE,nmax=5000,outdir=outDir)
28
29 # set location of mask for clipping rasters (substrate, V and h).
     Make sure its coordinate system is the same as for all other
     spatial data
30 modelBoundLoc<-"...\\hydromStress\\input\\clip.shp"
31
32 # set location of bounding box for seeding points (used to avoid
     points too close to the model's downstream boundary, which may
     have journeys that
33 # end outside the model
34 journeyMatrixBoxLoc<-"...\\hydromStress\\input\\seedPtsBox.shp"
35
36 # analysis parameters
37 t<-1459      # number of last time step for analyses (total time -
     1)
38 Delta<-10    # interval for area loss analysis (days)
39 lambda<-1    # decay rate for dispersal kerel
40 alpha<-1/10   # set alpha for target species
41 genTime<-365  # generation time of the species, in days
42 intervC<-7    # interval for calculation of c (connectivity)
43 iniAnim<-0    # initial time step for animation
44 finAnim<-365  # final time step for animation
45
46
47 # projection data (proj4 string)
48 projData<-"+proj=tmerc +lat_0=0 +lon_0=12 +k=1 +x_0=4500000 +y_0=0
     +ellps=krass +units=m
49   +a=6378245 +rf=298.3 +towgs84=28.000,-130.000,-95.000 +to_meter=1
```

"

patchMaps.R

```
# set environment
options(max.print=5.5E8,width=10000,signif=50) # set console
    options
library(akima)
library(raster)
library(sp)
library(maptools)
library(rgeos)
library(shapefiles)
library(animation)
library(lattice)
library(igraph)
library(rgdal)

# Cycle for (linear) interpolation of velocity (V) and water depth
    (h) onto a regular grid
# Creates a SpatialPolygons object and writes it to an ESRI
    shapefile for each variable at each time step
# Projection data must be provided in control script (variable
    projData)

# read velocity field, water depth and substrate map
v<-read.table(vData,header=TRUE,sep=",")
h<-read.table(hData,header=TRUE,sep=",")
sRast<-raster(readAsciiGrid(substrateRaster))

# read model boundary shapefile
modelBound<-readShapeSpatial(modelBoundLoc)

for (i in 3:(t+3)){

  # interpolate V and h using extent and resolution of substrate
      raster
  vInterp <- interp(x=v$x, y=v$y, z=v[[i+3]], yo=seq(from=extent(
      sRast)@ymin+res(sRast)[2]/2,
    length=nrow(sRast),by=res(sRast)[2]),xo=seq(from=extent(sRast)
        @xmin+res(sRast)[1]/2,
    length=ncol(sRast),by=res(sRast)[1]))

  hInterp <- interp(x=(h$x), y=h$y, z=h[[i+3]], yo=seq(from=extent(
      sRast)@ymin+res(sRast)[2]/2,
    length=nrow(sRast),by=res(sRast)[2]),xo=seq(from=extent(sRast)
        @xmin+res(sRast)[1]/2,
    length=ncol(sRast),by=res(sRast)[1]))

  # convert interpolations to raster objects
  vRast<-raster(vInterp)
  hRast<-raster(hInterp)

  # clip V and h rasters with mask; this step is very memory hungry
      !!
  vRast<-mask(vRast,modelBound)
  hRast<-mask(hRast,modelBound)
```

```
44
45   # overlay sRast, hRast and vRast to produce habitat suitability
       raster (HSrast, 1=habitable pixel, 0=unsuitable pixel)
46   HSrast<-(hRast>0.1 & vRast<0.6 & sRast==1)
47
48   # patch recognition step
49   HSrast[HSrast!=1]<-NA # leave only patch pixels, everything else
       is NA
50   HSclump<-clump(HSrast)  # clump pixels belonging to same patch (
       Moore neighborhood)
51   patches<-rasterToPolygons(HSclump,dissolve = F,digits=4)  #
       convert clump raster into vector polygons
52   patchesU<-unionSpatialPolygons(patches,patches@data[[1]]) # merge
        polygons according to patch membership
53
54   patchesU@proj4string<-CRS(projData)  # assign geographic
       projection
55
56   # extract IDs of cells making up habitat polygons (patches) and
       write them to SpatialPolygonsDataFrame (patchesU)
57   patchCellID<-extract(HSclump,patchesU,cellnumbers=TRUE)
58   idDF<-data.frame(1:length(patchesU@polygons))
59   for (j in 1:length(patchesU@polygons)){
60     idDF[[1]][j]<-paste(as.character(patchCellID[[j]][,1]),collapse
         =",")
61   }
62   colnames(idDF)<-"cell_IDs"
63   patchesU<-SpatialPolygonsDataFrame(patchesU,idDF)
64
65   # write shapefile
66   shapefile(patchesU,filename=paste(c(outDir,"patchPoly",(i-3)),
       collapse=""),overwrite=T)
67   # write .prj file
68   showWKT(p4s=proj4string(patchesU), file = paste(c(outDir,"
       patchPoly",(i-3),".prj"),collapse=""), morphToESRI = TRUE)
69
70   flush.console()
71   print(paste("printed patch shapefile",(i-3)))
72 }
```

patchTrack.R

```
1  # patch tracking algorithm
2  #---------------------------
3
4  # create cellID raster using substrate raster as template (extent
     and cell size)
5  cellIDrast<-raster(readAsciiGrid(substrateRaster))
6  cellIDrast[!is.na(cellIDrast)]<-1:length(cellIDrast[!is.na(
     cellIDrast)])
7
8  signatureList<-list()
9  signatureStorage<-list()
10
11 # build signature structure of study reach
12
13 # initialize with first time step
```

```
14 p<-readShapeSpatial(paste(c(outDir,"patchPoly",0),collapse=""))
15
16 #convert sub-polygons into polygons, then re-build p
17 polygonList<-list();length(polygonList)<-0
18 for (j in 1:length(p@polygons)){
19   count<-0
20   while (count<length(p@polygons[[j]]@Polygons)){
21     length(polygonList)<-length(polygonList)+1
22     polygonList[[length(polygonList)]]<-p@polygons[[j]]@Polygons[[
          count+1]]
23     count<-count+1
24   }
25 }
26 polygonSList<-list();length(polygonSList)<-length(polygonList)
27 for (j in 1:length(polygonSList)){
28   polygonSList[[j]]<-Polygons(list(polygonList[[j]]),j)
29 }
30 p<-SpatialPolygons(polygonSList)
31
32 # re-assign polygon IDs to match numbering, necessary for using
     gCentroid
33 for (i in 1:length(p@polygons)){p@polygons[[i]]@ID<-as.character(i)
     }
34
35 patchSigList<-list()
36 extr<-extract(cellIDrast,p,cellnumbers=TRUE)
37 for (j in 1:length(extr)){
38   patchSigList[[j]]<-extr[[j]][,1]
39 }
40 signatureList[[1]]<-patchSigList
41 signatureStorage[[1]]<-patchSigList
42
43 # loop through all remaining time steps
44 for (i in 1:t){
45
46   # read habitat polygons
47   p<-readShapeSpatial(paste(c(outDir,"patchPoly",i),collapse=""))
48
49   #convert sub-polygons into polygons, then re-build p
50   polygonList<-list();length(polygonList)<-0
51   for (j in 1:length(p@polygons)){
52     count<-0
53     while (count<length(p@polygons[[j]]@Polygons)){
54       length(polygonList)<-length(polygonList)+1
55       polygonList[[length(polygonList)]]<-p@polygons[[j]]@Polygons
            [[count+1]]
56       count<-count+1
57     }
58   }
59   polygonSList<-list();length(polygonSList)<-length(polygonList)
60   for (j in 1:length(polygonSList)){
61     polygonSList[[j]]<-Polygons(list(polygonList[[j]]),j)
62   }
63   p<-SpatialPolygons(polygonSList)
64
65   # re-assign polygon IDs to match numbering, necessary for using
```

```
      gCentroid
66    for (j in 1:length(p@polygons)){p@polygons[[j]]@ID<-as.character(
        j)}
67
68    # extract cell signatures and put them in list
69    patchSigList<-list()
70    extr<-extract(cellIDrast,p,cellnumbers=TRUE)
71    for (j in 1:length(extr)){
72      patchSigList[[j]]<-extr[[j]][,1]
73    }
74    signatureStorage[[i+1]]<-patchSigList
75    signatureList[[i+1]]<-list();length(signatureList[[i+1]])<-length
        (signatureList[[i]])
76
77    for (j in 1:length(patchSigList)){
78      marker<-0
79      for (k in 1:i){
80        for (l in 1:length(signatureList[[k]])){
81          if(length(intersect(patchSigList[[j]],signatureList[[k]][[l
              ]]))>0){
82            signatureList[[i+1]][[l]]<-patchSigList[[j]]
83            marker<-1
84            break}
85        }
86        if (marker==1) break
87      }
88      if (marker==0){
89        length(signatureList[[i+1]])<-length(signatureList[[i+1]])+1
90        signatureList[[i+1]][[length(signatureList[[i+1]])]]<-
            patchSigList[[j]]
91        lengthUpdate<-numeric()
92        for (m in 1:length(signatureList)){
93          lengthUpdate[m]<-lapply(signatureList,length)[[m]]
94        }
95        for (m in 1:length(signatureList)){
96          length(signatureList[[m]])<-max(lengthUpdate)
97        }}
98    }
99
100   flush.console()
101   print(paste("building signature structure, step ",i+1,"/",t+1))
102 }
103
104 # build areasMatrix, xCentroidMatrix and yCentroidMatrix
105 areasMatrix<-matrix(ncol=lapply(signatureList,length)[[1]],nrow=t
      +1)
106 xCentroidMatrix<-matrix(ncol=lapply(signatureList,length)[[1]],nrow
      =t+1)
107 yCentroidMatrix<-matrix(ncol=lapply(signatureList,length)[[1]],nrow
      =t+1)
108
109 for (i in 0:t){
110   p<-readShapeSpatial(paste(c(outDir,"patchPoly",i),collapse=""))
111
112   # re-assign polygon IDs to match numbering, necessary for using
        gCentroid
```

```
113   for (j in 1:length(p@polygons)){p@polygons[[j]]@ID<-as.character(
        j)}
114
115   for (j in 1:length(signatureStorage[[i+1]])){
116     for (k in 1:length(signatureList[[i+1]])){
117       if(length(intersect(signatureStorage[[i+1]][[j]],
            signatureList[[i+1]][[k]]))>0){
118         areasMatrix[i+1,k]<-p@polygons[[j]]@area
119         xCentroidMatrix[i+1,k]<-gCentroid(p,byid=TRUE)@coords[j,1]
120         yCentroidMatrix[i+1,k]<-gCentroid(p,byid=TRUE)@coords[j,2]
121         break}
122     }
123   }
124
125   flush.console()
126   print(paste("building areasMatrix, step ",i+1,"/",t+1))
127 }
128
129 #replace NAs with zero
130 areasMatrix[is.na(areasMatrix)]<-0
131 xCentroidMatrix[is.na(xCentroidMatrix)]<-0
132 yCentroidMatrix[is.na(yCentroidMatrix)]<-0
133
134 # write text files with output matrices
135 write.table(areasMatrix, file = paste0(outDir,"areasMatrix.txt"),
      quote=FALSE,
136   sep=",",row.names=FALSE,col.names=FALSE)
137 write.table(xCentroidMatrix, file = paste0(outDir,"xCentroidMatrix.
      txt"),quote=FALSE,
138   sep=",",row.names=FALSE,col.names=FALSE)
139 write.table(yCentroidMatrix, file = paste0(outDir,"yCentroidMatrix.
      txt"),quote=FALSE,
140   sep=",",row.names=FALSE,col.names=FALSE)
```

ind_area_analysis.R

```
 1 # area and area loss analyses
 2
 3 # read data
 4 areasMatrix<-as.matrix(read.table(paste0(outDir,"areasMatrix.txt",
 5   collapse=""),sep=",",header=F));colnames(areasMatrix)<-NULL
 6 xCentroidMatrix<-as.matrix(read.table(paste0(outDir,"
      xCentroidMatrix.txt",
 7   collapse=""),sep=",",header=F));colnames(xCentroidMatrix)<-NULL
 8 yCentroidMatrix<-as.matrix(read.table(paste0(outDir,"
      yCentroidMatrix.txt",
 9   collapse=""),sep=",",header=F));colnames(yCentroidMatrix)<-NULL
10
11 # total habitable area
12 A<-numeric()
13 for (i in 1:nrow(areasMatrix)){A[i]<-sum(areasMatrix[i,])}
14
15 # number of patches
16 N<-numeric()
17 for (i in 1:nrow(areasMatrix)){N[i]<-length(areasMatrix[i,][
      areasMatrix[i,]!=0])}
18
```

```
19 # patch-specific area losses
20 lossList<-list()
21 for (j in 1:ncol(areasMatrix)){
22   l_i<-numeric()
23   for (i in 1:(nrow(areasMatrix)-Delta)){
24     l_i[i]<-(areasMatrix[i+Delta,j]-areasMatrix[i,j])/areasMatrix[i
         ,j]
25 
26   }
27   l_i<-l_i[!is.na(l_i) & is.finite(l_i) & l_i<0]
28   lossList[[j]]<-(-l_i)
29 
30   flush.console()
31   print(j)
32 }
33 
34 # whole-network area losses
35 l_Delta<-numeric()
36 for (i in 1:(nrow(areasMatrix)-Delta)){
37   diff<-areasMatrix[i+Delta,]-areasMatrix[i,]
38 
39   #divide size of area losses between t and t+Delta by total
       available
40   # area at time t
41   l_Delta[i]<-sum(abs(diff[which(diff<0)]))/sum(areasMatrix[i,])
42 
43   flush.console()
44   print(i)
45 }
46 
47 # spatial distribution of l_i (=patch-specific l_i)
48 
49 # calculate average value of X and Y centroid coordinates
50 xCbar<-numeric()
51 for (i in 1:ncol(xCentroidMatrix)){
52   xCbar[i]<-mean(xCentroidMatrix[,i][xCentroidMatrix[,i]>0])
53 }
54 yCbar<-numeric()
55 for (i in 1:ncol(yCentroidMatrix)){
56   yCbar[i]<-mean(yCentroidMatrix[,i][yCentroidMatrix[,i]>0])
57 }
58 
59 #turn into SpatialPointsDataFrame object
60 pr_li_le.5<-numeric()
61 for (i in 1:length(lossList)){
62   pr_li_le.5[i]<-length(lossList[[i]][lossList[[i]]<.5])/
63     length(lossList[[i]])
64 }
65 XYc_bar<-SpatialPointsDataFrame(coords=cbind(xCbar,yCbar),
66   data=data.frame(pr_li_le.5))
67 
68 # write text files with results
69 write(N,paste0(outDir,"N.txt",collapse=""),sep="\n")
70 write(A,paste0(outDir,"A.txt",collapse=""),sep="\n")
71 sink(paste0(outDir,"lossList.txt",collapse=""));lossList;sink()
72 write(l_Delta,paste0(outDir,"l_Delta.txt",collapse=""),sep="\n")
```

```
73 sink(paste0(outDir,"liSpatial.txt",collapse=""));data.frame(XYc_bar
     );sink()
```

ind_patch_duration.R

```
1  # patch duration analysis; detecting and quantifying patch
      existence spells
2
3  areasMatrix<-as.matrix(read.table(paste0(outDir,"areasMatrix.txt",
4    collapse=""),sep=",",header=F));colnames(areasMatrix)<-NULL
5  xCentroidMatrix<-as.matrix(read.table(paste0(outDir,"
      xCentroidMatrix.txt",
6    collapse=""),sep=",",header=F));colnames(xCentroidMatrix)<-NULL
7  yCentroidMatrix<-as.matrix(read.table(paste0(outDir,"
      yCentroidMatrix.txt",
8    collapse=""),sep=",",header=F));colnames(yCentroidMatrix)<-NULL
9
10
11 # build individual patch durations
12 patchDurationList<-list()
13 for (i in 1:ncol(areasMatrix)){
14   exMatrix<-matrix(nrow=1,ncol=2)
15   if (areasMatrix[,i][length(areasMatrix[,i])]!=0){
16     # build working vector
17     wV<-c(0,which(areasMatrix[,i]==0),nrow(areasMatrix))
18     for (j in 1:(length(wV)-1)){
19       patchDur<-wV[j+1]-wV[j]-1
20       if (patchDur>0){
21         exMatrix<-rbind(exMatrix,as.vector(c(wV[j]+1,
22           patchDur)))
23       }
24     }
25     exMatrix<-cbind(exMatrix[2:nrow(exMatrix),1],
26       exMatrix[2:nrow(exMatrix),2])
27     colnames(exMatrix)<-c("start","duration")
28     patchDurationList[[i]]<-exMatrix
29   } else {
30     wV<-c(0,which(areasMatrix[,i]==0))
31     for (j in 1:(length(wV)-1)){
32       patchDur<-wV[j+1]-wV[j]-1
33       if (patchDur>0){
34         exMatrix<-rbind(exMatrix,as.vector(c(wV[j]+1,
35           patchDur)))
36       }
37     }
38   exMatrix<-cbind(exMatrix[2:nrow(exMatrix),1],exMatrix[2:nrow(
        exMatrix),
39     2])
40   colnames(exMatrix)<-c("start","duration")
41   patchDurationList[[i]]<-exMatrix
42   }
43
44   flush.console()
45   print(i)
46 }
47
48 # compute descriptors of duration distribution
```

```
49 totalDur<-numeric()
50 minDur<-numeric()
51 avgDur<-numeric()
52 maxDur<-numeric()
53 numberSpells<-numeric()
54 interSpellDurList<-list();length(interSpellDurList)<-length(
     patchDurationList)
55 minIntDur<-numeric()
56 avgIntDur<-numeric()
57 maxIntDur<-numeric()
58 medianIntDur<-numeric()
59 
60 for (i in 1:length(patchDurationList)){
61   totalDur[i]<-sum(patchDurationList[[i]][,2])
62   minDur[i]<-min(patchDurationList[[i]][,2])
63   avgDur[i]<-mean(patchDurationList[[i]][,2])
64   maxDur[i]<-max(patchDurationList[[i]][,2])
65   numberSpells[i]<-length(patchDurationList[[i]][,2])
66 
67   # inter-spell duration
68   # go to next i (patch) if present patch has only one duration
       spell
69   # no inter-spell duration can be computed
70   if (nrow(patchDurationList[[i]])<=1){
71     interSpellDurList[[i]]<-NA
72     next
73   }
74   # otherwise, do calculation
75   for (j in 1:(nrow(patchDurationList[[i]])-1)){
76     interSpellDurList[[i]][j]<-patchDurationList[[i]][j+1,
77       1]-sum(patchDurationList[[i]][j,])-1
78   }
79 
80   flush.console()
81   print(i)
82 }
83 
84 avgIntDur<-sapply(interSpellDurList,mean,simplify="vector")
85 minIntDur<-sapply(interSpellDurList,min,simplify="vector")
86 maxIntDur<-sapply(interSpellDurList,max,simplify="vector")
87 medianIntDur<-sapply(interSpellDurList,median,simplify="vector")
88 
89 # spatial distribution of u_i (=patch-specific u_i)
90 
91 # calculate average value of X and Y centroid coordinates
92 xCbar<-numeric()
93 for (i in 1:ncol(xCentroidMatrix)){
94   xCbar[i]<-mean(xCentroidMatrix[,i][xCentroidMatrix[,i]>0])
95 }
96 yCbar<-numeric()
97 for (i in 1:ncol(yCentroidMatrix)){
98   yCbar[i]<-mean(yCentroidMatrix[,i][yCentroidMatrix[,i]>0])
99 }
100 
101 #turn into SpatialPointsDataFrame object
102 totalDurXY<-SpatialPointsDataFrame(coords=cbind(xCbar,yCbar),
```

```
103   data=data.frame(totalDur))
104 avgDurXY<-SpatialPointsDataFrame(coords=cbind(xCbar,yCbar),
105   data=data.frame(avgDur))
106 numberSpellsXY<-SpatialPointsDataFrame(coords=cbind(xCbar,yCbar),
107   data=data.frame(numberSpells))
108 avgIntDurXY<-SpatialPointsDataFrame(coords=cbind(xCbar,yCbar),
109   data=data.frame(avgIntDur))
110
111 # write text files with results
112 sink(paste0(outDir,"patchDurationList.txt",collapse=""));
      patchDurationList;
113   sink()
114 sink(paste0(outDir,"interSpellDurList.txt",collapse=""));
      interSpellDurList;
115   sink()
116 sink(paste0(outDir,"patchDuration.txt",collapse=""));cbind(totalDur
      ,minDur,
117   avgDur,maxDur,numberSpells,avgIntDur,minIntDur,maxIntDur,
        medianIntDur);
118   sink()
119 sink(paste0(outDir,"uSpatial.txt",collapse=""));data.frame(xCbar,
      yCbar,
120   totalDur,avgDur,numberSpells,avgIntDur);sink()
```

ind_journeyTrack.R

```
1 # journeyTracking for connectivity analysis
2
3 # read flow direction and velocity magnitude data
4 fd<-read.table(flowDirData,header=TRUE,sep=",");fd[is.na(fd)]<-0
5 vMag<-read.table(vData,header=TRUE,sep=",")
6
7 # read substrate raster (used as template in interpolation)
8 sRast<-raster(readAsciiGrid(substrateRaster))
9
10 # read model boundary shapefile
11 modelBound<-readShapeSpatial(modelBoundLoc)
12
13 # load bounding box for seeding points (used to avoid points too
     close to the model's downstream boundary, which may have
     journeys that
14 # end outside the model
15 journeyMatrixBox<-readShapeSpatial(journeyMatrixBoxLoc,proj4string=
     CRS(as.character(projData)))
16
17 c<-data.frame(cbind(NA,NA))
18 colnames(c)[1]<-"t";colnames(c)[2]<-"c"
19 iter<-0 # counter for rows of c data frame
20
21 # this matrix summarizes the characteristics of the hypothetical
     journeys (starting point, source patch, contact with another
     patch? (0/1), and end patch)
22 journeyMatrix<-matrix(ncol=5,nrow=0);colnames(journeyMatrix)<-c("
     startX","startY","sourcePatch","touch?","endPatch")
23
24 for (i in seq(from=0,to=t,by=intervC)){
25   iter<-iter+1
```

```
26
27  # load patch map
28  p<-readShapeSpatial(paste(c(outDir,"patchPoly",i),collapse=""))
29
30  #convert sub-polygons into polygons, then re-build p
31  polygonList<-list();length(polygonList)<-0
32  for (j in 1:length(p@polygons)){
33    count<-0
34    while (count<length(p@polygons[[j]]@Polygons)){
35      length(polygonList)<-length(polygonList)+1
36      polygonList[[length(polygonList)]]<-p@polygons[[j]]@Polygons
          [[count+1]]
37      count<-count+1
38    }
39  }
40  polygonSList<-list();length(polygonSList)<-length(polygonList)
41  for (j in 1:length(polygonSList)){
42    polygonSList[[j]]<-Polygons(list(polygonList[[j]]),j)
43  }
44  p<-SpatialPolygons(polygonSList)
45
46  # build flow direction and vMag rasters
47  # interpolate
48  fdInterp <- interp(x=fd$x, y=fd$y, z=fd[[i+3]], yo=seq(from=
      extent(sRast)@ymin+res(sRast)[2]/2,
49    length=nrow(sRast),by=res(sRast)[2]),xo=seq(from=extent(sRast)
        @xmin+res(sRast)[1]/2,
50    length=ncol(sRast),by=res(sRast)[1]))
51
52  vMagInterp <- interp(x=vMag$x, y=vMag$y, z=vMag[[i+3]], yo=seq(
      from=extent(sRast)@ymin+res(sRast)[2]/2,
53    length=nrow(sRast),by=res(sRast)[2]),xo=seq(from=extent(sRast)
        @xmin+res(sRast)[1]/2,
54    length=ncol(sRast),by=res(sRast)[1]))
55
56  # convert interpolations to raster object
57  fdRast<-raster(fdInterp)
58  vMagRast<-raster(vMagInterp)
59
60  # clip V and h rasters with mask; this step is very memory hungry
      !!
61  fdRast<-mask(fdRast,modelBound)
62  vMagRast<-mask(vMagRast,modelBound)
63
64  vMagRast[is.na(vMagRast)]<-0
65
66  # generate random seeding points (journey starting points) inside
        all patches
67  for (j in 1:length(p@polygons)){ # iterate through all patches (
      Polygons objects in p)
68    polyCoords<-p@polygons[[j]]@Polygons[[1]]@coords
69    pOver<-numeric()
70    sDF2<-matrix(ncol=2,nrow=0)
71    while (nrow(sDF2)<10){
72      xRand<-runif(1,min=min(polyCoords[,1]),max=max(polyCoords
          [,1]))
```

```
73      yRand<-runif(1,min=min(polyCoords[,2]),max=max(polyCoords
           [,2]))
74      # select points that fall inside polygon
75      indPoly<-list();indPoly[[1]]<-p@polygons[[j]]@Polygons[[1]]
76      indPoly<-Polygons(indPoly,1);indPoly<-SpatialPolygons(list(
           indPoly))
77      pOver<-over(x=SpatialPoints(data.frame(xRand,yRand)),y=
           indPoly)
78      sDF<-data.frame(xRand,yRand,pOver)
79      sDF<-sDF[which(!is.na(sDF[[3]])),1:2]
80      sDF2<-rbind(sDF2,sDF)
81    }
82    sDF2[,3]<-rep(j,nrow(sDF2))
83    sDF2[,4]<-NA;sDF2[,5]<-NA
84    colnames(sDF2)<-c("startX","startY","sourcePatch","touch?","
         endPatch")
85    journeyMatrix<-rbind(journeyMatrix,as.matrix(sDF2)) # grow
         journeyMatrix
86  }
87  rownames(journeyMatrix)<-1:nrow(journeyMatrix)
88
89  # clip journeyMatrix with bounding box
90  journeyMatrixCoords<-SpatialPoints(journeyMatrix[,1:2],
       proj4string=CRS(as.character(projData)))
91  journeyMatrixOver<-over(x=journeyMatrixCoords,y=journeyMatrixBox)
92  journeyMatrix<-cbind(journeyMatrix,journeyMatrixOver)
93  journeyMatrix<-journeyMatrix[which(!is.na(journeyMatrix[[6]]))
       ,1:5]
94
95  # calculate journeys and overlay them to patch map
96  for (j in 1:nrow(journeyMatrix)){ # do for all seed points
97    journeyMax<-rexp(1,rate=lambda) # draw random distance from
         dispersal kernel with decay rate = lambda (in control file)
98    journeyX<-numeric()
99    journeyX[1]<-as.vector(journeyMatrix[j,1])
100   journeyY<-numeric()
101   journeyY[1]<-as.vector(journeyMatrix[j,2])
102   # calculate journey
103   journeyLength<-0
104   l<-1
105   while (journeyMax>journeyLength){
106     if (extract(vMagRast,matrix(c(journeyX[l],journeyY[l]),ncol
           =2,nrow=1))<0.001){break}
107     journeyX[l+1]<-journeyX[l]+.3*cos(extract(fdRast,matrix(c(
           journeyX[l],journeyY[l]),ncol=2,nrow=1))) # 0.3 -> never
           skip a cell
108     journeyY[l+1]<-journeyY[l]+.3*sin(extract(fdRast,matrix(c(
           journeyX[l],journeyY[l]),ncol=2,nrow=1))) # 0.3 -> never
           skip a cell
109     journeyLength<-journeyLength+sqrt((journeyX[l+1]-journeyX[l])
           ^2+(journeyY[l+1]-journeyY[l])^2)
110     l<-l+1
111   }
112   journeyPts<-cbind(journeyX,journeyY)
113   # check whether the journey either ends within a patch (can
         also be source patch) or touches another patch
```

```
114     # this approach considers zero-length journeys (i.e., journeys
           starting at a point where vMag<0.001 m/s) as not
           contributing to connectivity
115     journeyMatrix[j,5]<-over(x=SpatialPoints(data.frame(journeyPts)
           [nrow(journeyPts),]),y=p) # find jourey endpoint and write
           it in journeyMatrix
116     # overlay journey (SpatialPoints object) onto patches. The
           resulting integer vector gives the IDs of the patches the
           journey touches. If the journey
117     # touches patches other than its source patch, it contributes
           to c
118     journeyTouch<-over(x=SpatialPoints(data.frame(journeyPts)),y=p)
119     journeyTouch<-journeyTouch[journeyTouch!=journeyMatrix[j,3] & !
           is.na(journeyTouch)]
120     if (length(journeyTouch)>0) journeyMatrix[j,4]<-1 else
           journeyMatrix[j,4]<-0
121   }
122   c[iter,1]<-i;c[iter,2]<-nrow(journeyMatrix[(journeyMatrix[,4]==1
        | !is.na(journeyMatrix[,5])),])/nrow(journeyMatrix)
123
124   flush.console()
125   print(i)
126 }
127
128 sink(paste0(outDir,"c.txt",collapse=""));c;sink()
```

xsInterp.R

```
1 # this algorithm densifies the input lines connecting the points in
      consecutive cross sectional profiles. care must be taken to
     ensure
2 # that the densification length is shorter than the shortest line
     segment in the input lines
3
4 # set console options and load libraries
5 setwd("c:\\doktor\\hydromStress\\scripts\\")
6 options(max.print=5.5E8,width=10000,signif=50)
7 library(akima)
8 library(raster)
9 library(sp)
10 library(maptools)
11 library(rgeos)
12 library(shapefiles)
13 library(animation)
14 library(lattice)
15 library(igraph)
16 library(rgdal)
17
18 # load data
19 xyz<-read.table("r_xyz.txt",header=TRUE,sep=",")
20 l<-readShapeSpatial("xsConn.shp")
21
22 # define empty objects
23 pLL<-list();length(pLL)<-nrow(xyz)
24 xyzInterp<-matrix(ncol=3,nrow=0)
25 cMll<-list()
26 # densify interpolation lines (l) (add equidistant points for all
```

```
    line segments longer than "dMax")
27 pb<-txtProgressBar(min = 0, max = length(l@lines),width = 20,style
    = 3)
28 dMax<-0.5
29 for (i in 1:length(l@lines)){
30   cM<-l@lines[[i]]@Lines[[1]]@coords
31   j<-1
32   while (j<nrow(cM)){
33     ipd<-sqrt((cM[j+1,1]-cM[j,1])^2+(cM[j+1,2]-cM[j,2])^2)
34     if (ipd>dMax){
35       nseg<-ceiling(ipd)*2
36       insRow<-matrix(nrow=nseg-1,ncol=2)
37       # calculate coordinates of new points
38       for (k in 1:nrow(insRow)){
39         insRow[k,1]<-cM[j,1]+k*(cM[j+1,1]-cM[j,1])/nseg
40         insRow[k,2]<-cM[j,2]+k*(cM[j+1,2]-cM[j,2])/nseg
41       }
42       # add rows to coordinate matrix between the current points
43       splitU<-cM[1:j,]
44       splitD<-cM[(j+1):(nrow(cM)),]
45       cM<-rbind(splitU,insRow,splitD,deparse.level=0)
46       j<-j+nseg
47     } else {
48       j<-j+1
49     }
50   }
51 
52   cMl<-list();cMl[[1]]<-Line(cM)
53   cML<-Lines(cMl,i-1)
54   cMll[[i]]<-cML
55   setTxtProgressBar(pb,i)
56 }
57 close(pb)
58 l<-SpatialLines(cMll)
59 # create squares centered at each xy point in the available
     elevation data points
60 for (i in 1:nrow(xyz)){
61   ul<-cbind(xyz[i,1]-.1,xyz[i,2]+.1)
62   ur<-cbind(xyz[i,1]+.1,xyz[i,2]+.1)
63   lr<-cbind(xyz[i,1]+.1,xyz[i,2]-.1)
64   ll<-cbind(xyz[i,1]-.1,xyz[i,2]-.1)
65   p<-Polygon(rbind(ul,ur,lr,ll,ul))
66   pl<-list();pl[[1]]<-p
67   pL<-Polygons(pl,i)
68   pLL[[i]]<-pL
69 }
70 xyzQ<-SpatialPolygonsDataFrame(SpatialPolygons(pLL),data=data.frame
     (xyz[[3]]))
71 # do spatial overlay to get z-coordinate of arc endpoints
72 pb<-txtProgressBar(min = 0, max = length(l@lines),width = 20,style
    = 3)
73 for (i in 1:length(l@lines)){
74   lSP<-SpatialPoints(l@lines[[i]]@Lines[[1]]@coords)
75   lz<-over(lSP,xyzQ)[[1]]
76   lxy<-l@lines[[i]]@Lines[[1]]@coords
77   lxyz<-cbind(lxy,lz)
```

```
78    # linear interpolation of z-values along line based on
         accumulated arc length percentage
79    # calculate accumulated arc length (linearly) point by point
80    d<-numeric()
81    for (j in 1:(nrow(lxyz)-1)){
82      d[j]<-sqrt((lxyz[j+1,1]-lxyz[j,1])^2+(lxyz[j+1,2]-lxyz[j,2])^2)
83    }
84    d<-d/sum(d)
85    if (lxyz[1,3]>lxyz[nrow(lxyz),3]) {u<- -1} else {u<- 1} # up or
         down?
86    dz<-u*d*abs(lxyz[1,3]-lxyz[nrow(lxyz),3])
87    for (j in 2:(nrow(lxyz)-1)){
88      lxyz[j,3]<-lxyz[j-1,3]+dz[j-1]
89    }
90    # get rid of first and last point (to avoid having more than one
         z-value at a single xy position)
91    lxyz<-lxyz[2:(nrow(lxyz)-1),]
92    xyzInterp<-rbind(xyzInterp,lxyz)
93    setTxtProgressBar(pb,i)
94  }
95  close(pb)
96  rownames(xyzInterp)<-NULL
97  write.table(xyzInterp,file="xyzInt.txt",quote=FALSE,sep="\t",
98    row.names=FALSE,col.names=c("x","y","z"))
```

12 REFERENCES

Addicott, J., Aho, J., Antolin, M., Padilla, D., Richardson, J., and Soluk, D. (1987). Ecological neighborhoods: scaling environmental patterns. *Oikos*, 49:340–346.

Akçakaya, H. R. (2000). Viability analyses with habitat-based metapopulation models. *Population Ecology*, 42:45–53.

Akima, H., Gebhardt, A., Petzoldt, T., and Maechler, M. (2013). *akima: Interpolation of irregularly spaced data.*

Allan, J. and Castillo, M. (2007). *Stream Ecology: Structure and function of running waters.* Springer.

Anderson, K. E., Paul, A. J., McCauley, E., Jackson, L. J., Post, J. R., and Nisbet, R. M. (2006). Instream flow needs in streams and rivers: the importance of understanding ecological dynamics. *Frontiers in Ecology and the Environment*, 4:309–318.

Angermeier, P. and Karr, J. (1994). Biological integrity versus biological diversity as policy directives. *Bioscience*, 44:690–697.

Arthington, A., James, C., Mackay, S., Rolls, R., Sternberg, D., and Barnes, A. (2012). Hydro-ecological relationships and thresholds to inform environmental flow management, Science Report, International WaterCentre, Brisbane.

Baptist, M. (2005). *Modelling Floodplain Biogeomorphology.* DUP Science.

Bates, P. and Roo, A. D. (2000). A simple raster-based model for flood inundation simulation. *Journal of Hydrology*, 236(1–2):54 – 77.

Bates, P. D., Horritt, M. S., and Fewtrell, T. J. (2010). A simple inertial formulation of the shallow water equations for efficient two-dimensional flood inundation modelling. *Journal of Hydrology*, 387:33–45.

Beechie, T. J., Sear, D. A., Olden, J. D., Pess, G. R., Buffington, J. M., Moir, H., Roni, P., and Pollock, M. M. (2010). Process-based principles for restoring river ecosystems. *Bioscience*, 60:209–222.

Beisel, J.-N., Usseglio-Polatera, P., and Moreteau, J.-C. (2000). The spatial heterogeneity of a river bottom: a key factor determining macroinvertebrate communities. *Hydrobiologia*, 422-423:163–171.

Bernhardt, E. S., Palmer, M. A., Allan, J. D., Alexander, G., Barnas, K., Brooks, S., Carr, J., Clayton, S., Dahm, C., Follstad-Shah, J., Galat, D., Gloss, S., Goodwin, P., Hart, D., Hassett, B., Jenkinson, R., Katz, S., Kondolf, G. M., Lake, P. S., Lave, R., Meyer, J. L., O'Donnell, T. K., Pagano, L., Powell, B., and Sudduth, E. (2005). Synthesizing u.s. river restoration efforts. *Science*, 308:636–637.

Bezzola, G. (2002). *Fliesswiderstand und Sohlenstabilitaet natuerlicher Gerinne unter besonderer Beruecksichtigung des Einflusses der relativen Ueberdeckung*. Dissertation, Eidgenoessische Technische Hochschule Zuerich.

Bilton, D., Freeland, J., and Okamura, B. (2001). Dispersal in freshwater invertebrates. *Annual Review of Ecology and Systematics*, 32:159–181.

Bivand, R., Keitt, T., and Rowlingson, B. (2013). *rgdal: Bindings for the Geospatial Data Abstraction Library*. R package version 0.8-12.

Bivand, R. and Lewin-Koh, N. (2013). *maptools: Tools for reading and handling spatial objects*. R package version 0.8-27.

Bivand, R. and Rundel, C. (2013). *rgeos: Interface to Geometry Engine - Open Source (GEOS)*. R package version 0.3-2.

Bizjak, A., Brooke, J., Bunzel, K., Holubová, K., Ionescu, I., Irmer, U., Koller-Kreimel, V., Laguna, M., Marttunen, M., Mohaupt, V., Moroz, S., Muotka, J., Naumann, S., Olsson, H., Pedersen, T., Pelikan, B., Piet, O., Pirker, O., Pollard, P., Rast, G., Rawson, J., Steinar, S., Törner, A., van der Molen, D., Vial, I., Vogel, B., and von Keitz, S. (2006). Good practice in managing the ecological impacts of hydropower schemes, flood protection works and works designed to facilitate navigation under the Water Framework Directive. Technical Paper on WFD and Hydromorphology Version 4.2, 30th November 2006.

Bleyel, B. and Faulhaber, P. (2007). Begutachtung des Elbe-Abschnitts Coswig (Elbe-km 220,2 - 245,6) mit Hilfe eines 2D HN-Modells. Teil 1: Istzustandsanalyse bis zum Ausuferungsabfluss.

Bokulich, A. (2013). Explanatory models versus predictive models: Reduced complexity modeling in geomorphology. In Karakostas, V. and Dieks, D., editors, *EPSA11 Perspectives and Foundational Problems in Philosophy of Science, The European Philosophy of Science Association Proceedings 2*, pages 115–128. Springer International Publishing Switzerland.

Bond, N. R. and Lake, P. S. (2003). Local habitat restoration in streams: Constraints on the effectiveness of restoration for stream biota. *Ecological Management & Restoration*, 4:193–198.

Bond, N. R., Perry, G. L. W., and Downes, B. J. (2000). Dispersal of organisms in a patchy stream environment under different settlement scenarios. *Journal of Animal Ecology*, 69:608–619.

Bovee, K. (1982). *A guide to stream habitat analysis using the instream flow incremental methodology. Instream Flow Information Paper 12.* U.S.D.I. Fish and Wildlife Service, Office of Biological Services. FWS/OBS-82/26.

Boyce, M. (1992). Population viability analysis. *Annual Review of Ecology and Systematics*, 23:481–506.

Boyero, L. (2003). Multiscale patterns of spatial variation in stream macroinvertebrate communities. *Ecological Research*, 18:365–379.

Brasington, J. and Richards, K. (2007). Reduced-complexity, physically-based geomorphological modelling for catchment and river management. *Geomorphology*, 90:171 – 177.

Brown, J. H., Gillooly, J. F., Allen, A. P., Savage, V. M., and West, G. B. (2004). Toward a metabolic theory of ecology. *Ecology*, 85:1771–1789.

Bunn, S. E. and Arthington, A. H. (2002). Basic principles and ecological consequences of altered flow regimes for aquatic biodiversity. *Environmental management*, 30:492–507.

Caissie, D. and El-Jabi, N. (2003). Instream flow assessment: From holistic approaches to habitat modelling. *Canadian Water Resources Journal*, 28:173–183.

Calabrese, J. M. and Fagan, W. F. (2004). A comparison-shopper's guide to connectivity metrics. *Frontiers in Ecology and the Environment*, 2:529–536.

Capra, H., Breil, P., and Souchon, Y. (1995). A new tool to interpret magnitude and duration of fish habitat variations. *Regulated Rivers: Research & Management*, 10:281–289.

CEN (2004). European Standard EN 14614:2004 - Water quality — Guidance standard for assessing the hydromorphological features of rivers.

Clarke, S. J., Bruce-Burgess, L., and Wharton, G. (2003). Linking form and function: towards an eco-hydromorphic approach to sustainable river restoration. *Aquatic Conservation: Marine and Freshwater Ecosystems*, 13:439–450.

Cleveland, W. S. (1993). *Visualizing Data.* Hobart Press, Summit, New Jersey, U.S.A.

Colwell, R. K. (1974). Predictability, constancy, and contingency of periodic phenomena. *Ecology*, 55(5):1148–1153. doi: 10.2307/1940366; M3: doi: 10.2307/1940366; 07.

Coulthard, T. J., Hicks, D. M., and Wiel, M. J. V. D. (2007). Cellular modelling of river catchments and reaches: Advantages, limitations and prospects. *Geomorphology*, 90:192–207.

Coulthard, T. J., Kirkby, M., and Macklin, M. (1996). A cellular automaton landscape evolution model. In *Proceedings of the first International Conference on GeoComputation (Volume 1), School of Geography, University of Leeds.* R.J. Abrahart.

Coulthard, T. J. and Van De Wiel, M. J. (2012). Modelling river history and evolution. *Philosophical Transactions of the Royal Society of London A: Mathematical, Physical and Engineering Sciences*, 370:2123–2142.

Cremona, F., Planas, D., and Lucotte, M. (2008). Biomass and composition of macroinvertebrate communities associated with different types of macrophyte architectures and habitats in a large fluvial lake. *Fundamental and Applied Limnology*, 171:119–130.

Crowder, D. and Diplas, P. (2000). Using two-dimensional hydrodynamic models at scales of ecological importance. *Journal of Hydrology*, 230:172–191.

Csardi, G. and Nepusz, T. (2006). The igraph software package for complex network research. *InterJournal*, Complex Systems:1695.

Dahm, V., Kupilas, B., Rolauffs, P., Hering, D., Haase, P., Kappes, H., Leps, M., and Sundermann, A. (2014). Strategien zur Optimierung von Fließgewässer-Renaturierungsmaßnahmen und ihrer Erfolgskontrolle. Umweltbundesamt, Texte 43/2014.

D'Angelo, D., Gregory, S., Ashkenas, L., and Meyer, J. (1997). Physical and biological linkages within a stream geomorphic hierarchy: a modeling approach. *Journal of the North American Benthological Society*, 16:480–502.

Daniels, M. D. (2006). Distribution and dynamics of large woody debris and organic matter in a low-energy meandering stream. *Geomorphology*, 77:286–298.

de Almeida, G. A. M. and Bates, P. (2013). Applicability of the local inertial approximation of the shallow water equations to flood modeling. *Water Resources Research*, 49:4833–4844.

de Almeida, G. A. M., Bates, P., Freer, J. E., and Souvignet, M. (2012). Improving the stability of a simple formulation of the shallow water equations for 2-d flood modeling. *Water Resources Research*, 48:14pp.

Didderen, K. and Verdonschot, P. (2010). Biological processes of connectivity and metapopulation dynamics in aquatic ecosystem restoration. Deliverable 6.4-1, Project wiser.

Diehl, S., Anderson, K., and Nisbet, R. (2008). Population responses of drifting stream invertebrates to spatial environmental variability: an emerging conceptual framework. In Lancaster, J. and Briers, R., editors, *Aquatic insects: Challenges to populations*, pages 158–183. CAB International.

Downes, B. (2010). Back to the future: little-used tools and principles of scientific inference can help disentangle effects of multiple stressors on freshwater ecosystems. *Freshwater Biology*, 55:60–79.

Downes, B. and Keough, M. (1998). Scaling of colonization processes in streams: Parallels and lessons from marine hard substrata. *Australian Journal of Ecology*, 23:8–26.

Downes, B. and Reich, P. (2008). What is the spatial structure of stream insect populations? dispersal behaviour at different life-history stages. In Lancaster, J. and Briers, R., editors, *Aquatic insects: Challenges to populations*. CAB International.

Downes, B. J., Hindell, J. S., and Bond, N. R. (2000). What's in a site? variation in lotic macroinvertebrate density and diversity in a spatially replicated experiment. *Austral Ecology*, 25(2):128–139.

Downes, B. J. and Lancaster, J. (2010). Does dispersal control population densities in advection-dominated systems? A fresh look at critical assumptions and a direct test. *Journal of Animal Ecology*, 79:235–248.

Doyle, M. W. (2006). A heuristic model for potential geomorphic influences on trophic interactions in streams. *Geomorphology*, 77:235–248.

Dufour, S. and Piégay, H. (2009). From the myth of a lost paradise to targeted river restoration: forget natural references and focus on human benefits. *River Research and Applications*, 25:568–581.

Dunning, J., Danielson, B., and Pulliam, H. (1992). Ecological processes that affect populations in complex landscapes. *Oikos*, 65:169–175.

EC (2000). Directive 2000/60/EC of the European Parliament and of the Council of 23 October 2000 establishing a framework for Community action in the field of water policy.

EC (2003). CIS guidance document No. 12 - horizontal guidance on the role of wetlands in the water framework directive. Common Implementation Strategy for the Water Framework Directive (2000/60/EC).

EEA (2011). Hazardous substances in Europe's fresh and marine waters - an overview. EEA Technical report No. 8/2011.

EEA (2012a). EEA Water 2012 Report: thematic assessment on ecological and chemical status and pressures. Version: 2 (draft version).

EEA (2012b). Hydromorphological alterations and pressures in european rivers, lakes, transitional and coastal waters. thematic assessment for EEA water 2012 report.

Elliott, J. M. (1971). The distances travelled by drifting invertebrates in a lake district stream. *Oecologia*, 6:350–379.

Fisher, S. G., Heffernan, J. B., Sponseller, R. A., and Welter, J. R. (2007). Functional ecomorphology: feedbacks between form and function in fluvial landscape ecosystems. *Geomorphology*, 89:84–96.

Fonseca, D. and Hart, D. (2001). Colonization history masks habitat preferences in local distributions of stream insects. *Ecology*, 82(10):2897–2910.

Fore, L. S., Karr, J. T., and Wisseman, R. W. (1996). Assessing invertebrate responses to human activities: Evaluating alternative approaches. *Journal of the North American Benthological Society*, 15(2):212–231.

Frissell, C., Liss, W., Warren, C., and Hurley, M. (1986). A hierarchical framework for stream habitat classification: Viewing streams in a watershed context. *Environmental Management*, 10(2):199–214.

Fuchs, U. and Statzner, B. (1990). Time scales for the recovery potential of river communities after restoration: lessons to be learned from smaller streams. *Regulated Rivers: Research & Management*, 5:77–87.

Gerisch, M., Dziock, F., Schanowski, A., Ilg, C., and Henle, K. (2012). Community resilience following extreme disturbances: the response of ground beetles to a severe summer flood in a central european lowland stream. *River Research and Applications*, 28:81–92.

Gerrish, N. and Bristow, J. M. (1979). Macroinvertebrate associations with aquatic macrophytes and artificial substrates. *Journal of Great Lakes Research*, 5:69–72.

Giebel, H., Rosenzweig, S., and Schleuter, M. (2011). Ökologische Modellierungen für die Wasser- und Schifffahrtsverwaltung – das integrierte Flussauenmodell INFORM in seiner neuesten Fassung (Version 3).

Gini, C. (1997). Concentration and dependency ratios. *Rivista di Politica Economica*, 87:769 – 792.

Haase, P., Hering, D., Jähnig, S. C., Lorenz, A. W., and Sundermann, A. (2013). The impact of hydromorphological restoration on river ecological status: a comparison of fish, benthic invertebrates, and macrophytes. *Hydrobiologia*, 704:475–488.

Hanski, I. (1999a). Habitat connectivity, habitat continuity, and metapopulations in dynamic landscapes. *Oikos*, 87:209–219.

Hanski, I. (1999b). *Metapopulation Ecology*. Oxford University Press.

Hart, D. and Finelli, C. (1999). Physical-biological coupling in streams: the pervasive effects of flow on benthic organisms. *Annual Review of Ecology and Systematics*, 30:363–395.

Haybach, A., Schöll, F., König, B., and Kohmann, F. (2004). Use of biological traits for interpreting functional relationships in large rivers. *Limnologica - Ecology and Management of Inland Waters*, 34:451 – 459.

Heino, J., Louhi, P., and Muotka, T. (2004). Identifying the scales of variability in stream macroinvertebrate abundance, functional composition and assemblage structure. *Freshwater Biology*, 49:1230–1239.

Hering, D., Borja, A., Carstensen, J., Carvalho, L., Elliott, M., Feld, C. K., Heiskanen, A.-S., Johnson, R. K., Moe, J., Pont, D., Solheim, A. L., and van de Bund, W. (2010). The european water framework directive at the age of 10: A critical review of the achievements with recommendations for the future. *Science of The Total Environment*, 408:4007–4019.

Hering, D., Buffagni, A., Moog, O., Sandin, L., Sommerhäuser, M., Stubauer, I., Feld, C., Johnson, R., Pinto, P., Skoulikidis, N., Verdonschot, P., and Zahrádková, S. (2003). The development of a system to assess the ecological quality of streams based on macroinvertebrates – design of the sampling programme within the aqem project. *International Review of Hydrobiology*, 88:345–361.

Hering, D., Johnson, R., and Buffagni, A. (2006). Linking organism groups – major results and conclusions from the STAR project. *Hydrobiologia*, 566:109–113.

Hijmans, R. J. (2013). *raster: Geographic data analysis and modeling.* R package version 2.1-66.

Hoffmann, A. and Hering, D. (2000). Wood-associated macroinvertebrate fauna in central european streams. *International Review of Hydrobiology*, 85:25–48.

Hughes, J. (2007). Constraints on recovery: using molecular methods to study connectivity of aquatic biota in rivers and streams. *Freshwater Biology*, 52:616–631.

Humphries, P. (1996). Aquatic macrophytes, macroinvertebrate associations and water levels in a lowland tasmanian river. *Hydrobiologia*, 321:219–233.

Hunter, N., Bates, P., Horritt, M., and Wilson, M. (2007). Simple spatially-distributed models for predicting flood inundation: A review. *Geomorphology*, 90:208–225.

Hurlbert, S. (1984). Pseudoreplication and the design of ecological field experiments. *Ecological Monographs*, 54:187–211.

Hutchinson, E. (1957). Concluding remarks. *Cold Spring Harbor symposia on quantitative biology*, 22:415–427.

IMPRESS (2003). Guidance Document No. 3 - Guidance for the analysis of pressures and impacts in accordance with the Water Framework Directive.

Januschke, K., Jähnig, S. C., Lorenz, A. W., and Hering, D. (2014). Mountain river restoration measures and their success(ion): Effects on river morphology, local species pool, and functional composition of three organism groups. *Ecological Indicators*, 38:243 – 255.

Jähnig, S. and Lorenz, A. (2008). Substrate-specific macroinvertebrate diversity patterns following stream restoration. *Aquatic Sciences*, 70:292–303.

Jähnig, S., Sommerhäuser, M., and Hering, D. (2011). Fließgewässerrenaturierung heute: Zielsetzungen, Methodik und Effizienzkontrolle. In Jähnig, S., Hering, D., and Sommerhäuser, M., editors, *Limnologie Aktuell: Fließgewässerrenaturierung heute und morgen - EG-Wasserrahmenrichtlinie, Maßnahmen und Effizienzkontrolle*, pages 1–6. Schweizerbart'sche Verlagsbuchhandlung.

Jähnig, S. C., Brabec, K., Buffagni, A., Erba, S., Lorenz, A. W., Ofenböck, T., Verdonschot, P. F. M., and Hering, D. (2010). A comparative analysis of restoration measures and their effects on hydromorphology and benthic invertebrates in 26 central and southern european rivers. *Journal of Applied Ecology*, 47:671–680.

Jähnig, S. C., Brunzel, S., Gacek, S., Lorenz, A. W., and Hering, D. (2009). Effects of re-braiding measures on hydromorphology, floodplain vegetation, ground beetles and benthic invertebrates in mountain rivers. *Journal of Applied Ecology*, 46:406–416.

Jähnig, S. C., Lorenz, A., and Hering, D. (2008). Hydromorphological parameters indicating differences between single- and multiple-channel mountain rivers in germany, in relation to their modification and recovery. *Aquatic Conservation: Marine and Freshwater Ecosystems*, 18(7):1200–1216.

Johnson, W. C. (2002). Riparian vegetation diversity along regulated rivers: contribution of novel and relict habitats. *Freshwater Biology*, 47:749–759.

Jorde, K., Schneider, M., Peter, A., and Zoellner, F. (2001). Fuzzy based models for the evaluation of fish habitat quality and instream flow assessment. In *Proceedings of the 2001 International Symposium on Environmental Hydraulics, Tempe, Arizona, USA*. ISEH.

Junk, J. W., Bayley, P. B., and Sparks, R. E. (1989). The flood pulse concept in river floodplain systems. *Canadian Special Publications of Fisheries and Aquatic Sciences*, 106:110–127.

Kail, J. and Hering, D. (2009). The influence of adjacent stream reaches on the local ecological status of central european mountain streams. *River Research and Applications*, 25:537–550.

Kail, J., Wolter, C., Radinger, J., Fohrer, N., Guse, B., Hering, D., and Schröder, M. (2012). Mieter für frischrenovierte Wohnung gesucht - Modellierungsansatz zur Prognose der Habitatbedingungen und Besiedlung (renaturierter) Gewässerabschnitte. In *Abstractband Jahrestatung der Deuschen Gesselschaft für Limnologie, 2012*. Universität Koblenz-Landau, Bundesanstalt für Gewässerkunde.

Kareiva, P. (1990). Population dynamics in spatially complex environments: theory and data. *Philosophical Transactions of the Royal Society of London: Biological Sciences*, 330:175–190.

Karr, J. R. and Chu, E. W. (1999). *Restoring life in running waters - better biological monitoring.* Island Press.

Keough, M. J., Quinn, G. P., and King, A. (1993). Correlations between human collecting and intertidal mollusc populations on rocky shores. *Conservation Biology*, 7:pp. 378–390.

Kirchesch, V., Bergfeld, T., and Müller, D. (2006). Auswirkungen der Stauregelung auf den Stoffhaushalt und die Trophie von Flüssen. In Jähnig, S., Hering, D., and Sommerhäuser, M., editors, *Limnologie Aktuell: Fließgewässer-Renaturierung heute und morgen - EG-Wasserrahmenrichtlinie, Maßnahmen und Effizienzkontrolle.* Schweizerbart'sche Verlagsbuchhandlung.

Kleinwächter, M. and Schilling, K. (2013). Alternative river bank protections – an appropriate approach to improve river banks along waterways from an ecological point of view? In *Proceedings of the 8th Symposium for European Freshwater Sciences - SEFS 8.*

Koenker, R. (2015). *quantreg: Quantile Regression.* R package version 5.11.

Koenker, R. and Bassett Jr., G. (1978). Regression quantiles. *Econometrica*, 46:33 – 50.

Koenker, R. and Hallock, K. (2000). Paper prepared for the journal of economic perspectives "symposium on econometric tools". consulted online on 16.4.2015.

Kopecki, I. (2008). *Calculational approach to FST hemispheres for multiparametrical benthos habitat modelling.* Dissertation, Universität Stuttgart, Fakultaet Bau- und Umweltingenieurwissenschaften.

Kopecki, I. and Schneider, M. (2010). Handbuch für das Habitatsimulationsmodell CASiMiR - Modul: CASiMiR-Benthos.

Korfiatis, K. and Stamou, G. (1999). Habitat templets and the changing worldview of ecology. *Biology and Philosophy*, 14:375–393.

Kuhn, W. and Kleyer, W. (1999). Landschaftsanalyse. In Amler, K., editor, *Populationsbiologie in der Naturschutzpraxis. Isolation, Fflächenbedarf und Biotopansprüche von Pflanzen und Tieren.* Ulmer (Eugen).

Lake, P. S., Bond, N., and Reich, P. (2007). Linking ecological theory with stream restoration. *Freshwater Biology*, 52:597–615.

Lamouroux, N., Mérigoux, S., Capra, H., Dolédec, S., Jowett, I. G., and Statzner, B. (2010). The generality of abundance-environment relationships in microhabitats: a comment on Lancaster and Downes (2009). *River Research and Applications*, 26:915–920.

Lancaster, J. (2008). Movement and dispersion of insects in stream channels: What role does flow play? In Lancaster, J. and Briers, R., editors, *Aquatic insects: Challenges to populations.* CAB International.

Lancaster, J. and Belyea, L. R. (2006). Defining the limits to local density: alternative views of abundance–environment relationships. *Freshwater Biology*, 51:783–796.

Lancaster, J. and Downes, B. J. (2010a). Ecohydraulics needs to embrace ecology and sound science, and to avoid mathematical artefacts. *River Research and Applications*, 26:921–929.

Lancaster, J. and Downes, B. J. (2010b). Linking the hydraulic world of individual organisms to ecological processes: Putting ecology into ecohydraulics. *River Research and Applications*, 26:385–403.

Lancaster, J. and Downes, B. J. (2013). Aquatic entomology. Oxford, 296 pp.

Lancaster, J., Downes, B. J., and Glaister, A. (2009). Interacting environmental gradients, trade-offs and reversals in the abundance–environment relationships of stream insects: when flow is unimportant. *Marine and Freshwater Research*, 60:259–270.

Lancaster, J., Hildrew, A. G., and Gjerlov, C. (1996). Invertebrate drift and longitudinal transport processes in streams. *Canadian Journal of Fisheries and Aquatic Sciences*, 53:572–582.

Lane, S. and Ferguson, R. (2005). Modelling of reach-scale fluvial flows. In Bates, P., Lane, S., and Ferguson, R., editors, *Computational Fluid Dynamics: Applications in Environmental Hydraulics*. John Wiley & Sons, Ltd.

Latimer, A. M., Wu, S., Gelfand, A. E., and J.A., S. J. (2006). Building statistical models to analyze species distributions. *Ecological Applications*, 16:33–50.

LAWA (2000). Gewässerstrukturgütekartierung in der Bundesrepublik Deutschland – Verfahren für kleine und mittelgroße Fließgewässer. Empfehlung.

Leal, C. (2012). The effects of restored aquatic large woody debris structures on invertebrate populations in the Napa river. Master's thesis, San Jose State University.

Leclerc, M. (2005). Ecohydraulics: a new interdisciplinary frontier for CFD. In Bates, P., Lane, S., and Ferguson, R., editors, *Computational Fluid Dynamics: Applications in Environmental Hydraulics*. John Wiley & Sons, Ltd.

Leibold, M. A., Holyoak, M., Mouquet, N., Amarasekare, P., Chase, J. M., Hoopes, M. F., Holt, R. D., Shurin, J. B., Law, R., Tilman, D., Loreau, M., and Gonzalez, A. (2004). The metacommunity concept: a framework for multi-scale community ecology. *Ecology Letters*, 7:601–613.

Levins, R. (1962). Theory of Fitness in a Heterogeneous Environment. I. The Fitness Set and Adaptive Function. *The American Naturalist*, 96:361–373.

Li, W., Han, R., Chen, Q., Qu, S., and Cheng, Z. (2010). Individual-based modelling of fish population dynamics in the river downstream under flow regulation. *Ecological Informatics*, 5(2):115 – 123.

Liebhold, A. M. and Gurevitch, J. (2002). Integrating the statistical analysis of spatial data in ecology. *Ecography*, 25:553–557.

Liess, M. and von der Ohe, P. (2005). Analyzing effects of pesticides on invertebrate communities in streams. *Environmental Toxicology and Chemistry*, 24:954–965.

Lorenz, A., Kirchner, L., and Hering, D. (2004). Electronic subsampling of macrobenthic samples: how many individuals are needed for a valid assessment result? *Hydrobiologia*, 516:299–312.

Macneale, K. H., Peckarsky, B. L., and Likens, G. E. (2005). Stable isotopes identify dispersal patterns of stonefly populations living along stream corridors. *Freshwater Biology*, 50:1117–1130.

Marcus, W. A. and Fonstad, M. A. (2008). Optical remote mapping of rivers at sub-meter resolutions and watershed extents. *Earth Surface Processes and Landforms*, 33:4–24.

Meier, C., Böhmer, J., Biss, R., Feld, C., Haase, P., Lorenz, A., Rawer-Jost, C., Rolauffs, P., Schindehütte, K., Schöll, F., Sundermann, A., Zenker, A., and Hering, D. (2006). Weiterentwicklung und Anpassung des nationalen Bewertungssystems für Makrozoobenthos an neue internationale Vorgaben. Umweltforschungsplan des Bundesministeriums für Umwelt, Naturschutz und Reaktorsicherheit, Förderkennzeichen (UFOPLAN) 202 24 223. 198pp.

Minshall, W. and Minshall, J. (1977). Microdistribution of benthic invertebrates in a rocky mountain (u.s.a) stream. *Hydrobiologia*, 55:231–249.

Moog, O. (2002). *Fauna aquatica Austriaca. A comprehensive species inventory of Austrian aquatic organisms with ecological notes*. Federal Ministry of Agriculture, Forestry, Environment and Water Management, Division VII (Water), Vienna, Austria, 2nd edition.

Moore, R., Spittlehouse, D., and Story, A. (2005). Riparian microclimate and stream temperature response to forest harvesting: a review. *Journal of the American Water Resources Association*, 41:813–834.

Murray, A. B. and Paola, C. (1994). A cellular model of braided rivers. *Nature*, 371:54–57.

Neal, J., Villanueva, I., Wright, N., Willis, T., Fewtrell, T., and Bates, P. (2012). How much physical complexity is needed to model flood inundation? *Hydrological Processes*, 26:2264–2282.

Newbold, J. D., Elwood, J. W., O'Neill, R. V., and Winkle, W. V. (1981). Measuring nutrient spiralling in streams. *Canadian Journal of Fisheries and Aquatic Sciences*, 38:860–863.

Newson, M. D. and Large, A. R. G. (2006). 'Natural' rivers, 'hydromorphological quality' and river restoration: a challenging new agenda for applied fluvial geomorphology. *Earth Surface Processes and Landforms*, 31:1606–1624.

Nicholas, A. P. (2005). Cellular modelling in fluvial geomorphology. *Earth Surface Processes and Landforms*, 30:645–649.

Nicholas, A. P. (2009). Reduced-complexity flow routing models for sinuous single-thread channels: intercomparison with a physically-based shallow-water equation model. *Earth Surface Processes and Landforms*, 34:641–653.

Nikora, V. (2010). Hydrodynamics of aquatic ecosystems: an interface between ecology, biomechanics and environmental fluid mechanics. *River Research and Applications*, 26:367–384.

Nikora, V., McEwan, I., McLean, S., Coleman, S., Pokrajac, D., and Walters, R. (2007). Double-averaging concept for rough-bed open-channel and overland flows: theoretical background. *Journal of Hydraulic Engineering*, 133:873–883.

Nilsson, C., Ekblad, A., Gardfjell, M., and Carlberg, B. (1991). Long-term effects of river regulation on river margin vegetation. *Journal of Applied Ecology*, 28:963–987.

Nilsson, C., Jansson, R., and Zinko, U. (1997). Long-term responses of river-margin vegetation to water-level regulation. *Science*, 276:798–800.

Nilsson, C., Reidy, C. A., Dynesius, M., and Revenga, C. (2005). Fragmentation and flow regulation of the world's large river systems. *Science*, 308:405–408.

ÖNORM-M-6232 (1997). Richtlinien für die ökologische Untersuchung und Bewertung von Fließgewässern.

O'Neill, R., Hunsaker, C., Timmins, S., Jackson, B., Jones, K., Riitters, K., and Wickham, J. (1996). Scale problems in reporting landscape pattern at the regional scale. *Landscape Ecology*, 11:169–180.

O'Neill, R. V. (1986). *A Hierarchical Concept of Ecosystems*. Princeton University Press.

O'Neill, R. V., Milne, B. T., Turner, M. G., and Gardner, R. H. (1988). Resource utilization scales and landscape pattern. *Landscape Ecology*, 2:63–69.

Orians, G. (1980). Micro and macro in ecological theory. *BioScience*, 30:79–79.

Osterkamp, W. R. and Hupp, C. R. (2010). Fluvial processes and vegetation — glimpses of the past, the present, and perhaps the future. *Geomorphology*, 116:274–285.

Palmer, M. A., Allan, J. D., and Butman, C. A. (1996). Dispersal as a regional process affecting the local dynamics of marine and stream benthic invertebrates. *Trends in Ecology & Evolution*, 11:322—326.

Palmer, M. A., Ambrose, R. F., and Poff, N. L. (1997). Ecological theory and community restoration ecology. *Restoration Ecology*, 5:291–300.

Palmer, M. A., Bernhardt, E. S., Allan, J. D., Lake, P. S., Alexander, G., Brooks, S., Carr, J., Clayton, S., Dahm, C. N., Shah, J. F., Galat, D. L., Loss, S. G., Goodwin, P., Hart, D. D., Hassett, B., Jenkinsin, R., Kondolf, G. M., Lave, R., Meyer, J. L., O'Donnell, T. K., Pagano, L., and Sudduth, E. (2005). Standards for ecologically successful river restoration. *Journal of Applied Ecology*, 42:208–217.

Palmer, M. A., Menninger, H. L., and Bernhardt, E. Y. (2010). River restoration, habitat heterogeneity and biodiversity: a failure of theory or practice? *Freshwater Biology*, 55:205–222.

Parasiewicz, P. (2007). Using mesoHABSIM to develop reference habitat template and ecological management scenarios. *River Research and Applications*, 23:924–932.

Pebesma, E. J. and Bivand, R. S. (2005). Classes and methods for spatial data in R. *R News*, 5:9–13.

Peeters, E., Camu, J., Beijer, J., Scheffer, M., and Gardeniers, J. (2002). Response of the waterlouse asellus aquaticus to multiple stressors: effects of current velocity and mineral substratum. *Journal of Aquatic Ecosystem Stress and Recovery*, 9:193–203.

Petersen, I., Masters, Z., Hildrew, A., and Ormerod, S. (2004). Dispersal of adult aquatic insects in catchments of differing land use. *Journal of Applied Ecology*, 41:934–950.

Pickett, S., Kolasa, J., and Jones, C. (2007). *Ecological Understanding.* Academic Press.

Piñeiro, G., Perelman, S., Guerschman, J. P., and Paruelo, J. M. (2008). How to evaluate models: Observed vs. predicted or predicted vs. observed? *Ecological Modelling*, 216:316 – 322.

Pitt, D. and Batzer, D. (2011). Woody debris as a resource for aquatic macroinvertebrates in stream and river habitats of the southeastern united states: a review. In *Proceedings of the 2011 Georgia Water Resources Conference.*

Poff, L., Allan, D., Bain, M., Karr, J., Prestegaard, K., Richter, B., Sparks, R., and Stromberg, J. (1997). The natural flow regime. *Bioscience*, 47:769–784.

Poff, N. and Ward, J. (1990). Physical habitat template of lotic systems: recovery in the context of historical pattern of spatiotemporal heterogeneity. *Environmental Management*, 14:629–645.

Poff, N. L. R. (1997). Landscape filters and species traits: towards mechanistic understanding and prediction in stream ecology. *Journal of the North American Benthological Society*, 16:391–409.

Poole, G. (2000). *Analysis and Dynamic Simulation of Morphologic Controls on Surface- and Ground-water Flux in a Large Alluvial Flood Plain.* Dissertation, University of Montana, Missoula.

Poole, G. (2002). Fluvial landscape ecology: addressing uniqueness within the river discontinuum. *Freshwater Biology*, 47:641–660.

Pottgiesser, T., Kail, J., Halle, M., Mishke, U., Müller, A., Seuter, S., van der Weyer, K., and Wolter, C. (2008). Das gute ökologische Potenzial: Methodische Herleitung und Beschreibung - morphologische und biologische Entwicklungspotenziale der Landes- und Bundeswasserstraßen im Elbgebiet (Endbericht PEWA II).

Pottgiesser, T., Kail, J., Mishke, U., Wolter, C., Rehfeld-Klein, M., Köhler, A., and van der Weyer, K. (2009). Das gute ökologische Potenzial von Wasserstraßen. *KW - Korrespondenz Wasserwirtschaft*, 2:472–478.

Pottgiesser, T. and Sommerhäuser, M. (2008). Profiles of German stream types. Type 9: Mid-sized fine to coarse substrate dominated siliceous highland rivers.

Power, M., Sun, A., Parker, G., Dietrich, W., and Wootton, T. (1995). Hydraulic food-chain models - an approach to the study of food-web dynamics in large rivers. *Bioscience*, 45:159–167.

Pringle, C. M., Naiman, R. J., Bretschko, G., Karr, J. R., Oswood, W., Webster, J. R., Welcomme, R. L., and Winterbourn, M. J. (1988). Patch dynamics in lotic systems : the stream as a mosaic. *Journal of the North American Benthological Society*, 7:503–524.

Quinn, G. P. and Keough, M. J. (2002). *Experimental design and data analysis for biologists.* Cambridge University Press, Cambridge, UK, New York.

Ranta, E., Lundberg, P., and Kaitala, V. (2006). *Ecology of Populations*. Cambridge University Press, New York.

Raven, P., Holmes, N., Dawson, H., Fox, P., Everard, M., Fozzard, I., and Rouen, K. (1998). *River Habitat Quality: the Physical Character of Rivers and Streams in the UK and Isle of Man. River Habitat Survey Report No. 2.* Environment Agency of Grean Britain; Scottish Environment Protection Agency; Environment and Heritage Service of Northern Ireland.

Reuter, J. M., Jacobson, R. B., Elliott, C. M., III, H. E. J., and DeLonay, A. J. (2008). *Hydraulic and Substrate Maps of Reaches Used by Sturgeon (Genus Scaphirhynchus) in the Lower Missouri River, 2005–07: U.S. Geological Survey Data Series 386.* U.S. Geological Survey.

Rice, S. P., Buffin-Bélanger, T., Lancaster, J., and Reid, I. (2008). Movements of a macroinvertebrate (Potamophylax latipennis) across a gravel-bed substrate: effects of local hydraulics and micro-topography under increasing discharge. In Habersack, H., Piégay, H., and Rinaldi, M., editors, *Gravel-bed rivers VI: from process understanding to river restoration*, pages 637–659. Elsevier.

Rinaldi, M., Surian, N., Comiti, F., and Bussettini, M. (2013). A method for the assessment and analysis of the hydromorphological condition of Italian streams: The Morphological Quality Index (MQI). *Geomorphology*, 180-181:96.

Rochette, S., Huret, M., Rivot, E., and Pape, O. L. (2012). Coupling hydrodynamic and individual-based models to simulate longterm larval supply to coastal nursery areas. *Fisheries Oceanography*, 21:229–242.

Rowell, T. J. and Sobczak, W. V. (2008). Will stream periphyton respond to increases in light following forecasted regional Hemlock mortality? *Journal of Freshwater Ecology*, 23:33–40.

Sarkar, D. (2008). *Lattice: Multivariate Data Visualization with R.* Springer, New York.

Schmidt-Kloiber, A. and Hering, D. (2012). www.freshwaterecology.info - the taxa and autecology database for freshwater organisms, version 5.0.

Schneider, D. (1994). *Quantitative Ecology: Spatial and Temporal Scaling.* Academic Press.

Schneider, K. and Winemiller, K. (2008). Structural complexity of woody debris patches influences fish and macroinvertebrate species richness in a temperate floodplain-river system. *Hydrobiologia*, 610:235–244.

Schröder, M., Hering, D., Sondermann, M., Gies, M., and Feld, C. (2012). Identifizierung von Quellpopulationen in Fließgewässern: Entwicklfung eines Verbreitungsmodells für ausgewählte Fließgewässerorganismen in einem deutschen Tieflandgewässer (Treene) und Vergleich mit aktuellen Modellierungsansätzen aus dem Mittelgebirge. In *Abstractband Jahrestagung der Deutschen Gesellschaft für Limnologie, 2012.* Universität Koblenz-Landau, Bundesanstalt für Gewässerkunde.

Schröder, M., Kiesel, J., Schattmann, A., Jähnig, S. C., Lorenz, A. W., Kramm, S., Keizer-Vlek, H., Rolauffs, P., Graf, W., Leitner, P., and Hering, D. (2013). Substratum associations of benthic invertebrates in lowland and mountain streams. *Ecological Indicators*, 30(0):178 – 189.

Schuwirth, N. (2012). Predicting the occurrence of macroinvertebrates. *Eawag News*, 72:14–17.

Schuwirth, N. and Reichert, P. (2013). Bridging the gap between theoretical ecology and real ecosystems: modeling invertebrate community composition in streams. *Ecology*, 94:368–379.

Shaffer, M. (1981). Minimum population sizes for species conservation. *Bioscience*, 31:131–134.

Sharpe, A. and Downes, B. (2006). The effects of potential larval supply, settlement and post-settlement processes on the distribution of two species of filter-feeding caddisflies. *Freshwater Biology*, 51:717–729.

Shields Jr., D. (2010). Aquatic habitat bottom classification using ADCP. *Journal of Hydraulic Engineering*, 136(5):336–342.

Slobodkin, L. (1968). Toward a predictive theory of evolution. In Lewontin, R., editor, *Population biology and evolution*, pages 187–205. Syracuse University Press, Syracuse, New York.

Sober, E. (1985). *The Nature of Selection: Evolutionary Theory in Philosophical Focus*. MIT Press: Cambridge, Massachusetts.

Sondermann, M., Gies, M., Schröder, M., Hering, D., and Feld, C. (2012). Die Prognose der Wiedebesiedlung von Fließgewässern durch Quellpopulationen zweier Arten des Makrozoobenthos: konzeptionelle und technische Umsetzung in einem geographischen Informationssystem. In *Abstractband Jahrestatung der Deuschen Gesselschaft für Limnologie, 2012*. Universität Koblenz-Landau, Bundesanstalt für Gewässerkunde.

Southwood, T. (1977). Habitat, the templet for ecological strategies? *Journal of Animal Ecology*, 46:337–365.

Southwood, T. (1988). Tactics, strategies and templets. *Oikos*, 52:3–18.

Stabler, B. (2013). *shapefiles: Read and Write ESRI Shapefiles*. R package version 0.7.

Stalnaker, C., Lamb, B., Henriksen, J., Bovee, K., and Bartholow, J. . (1995). *The Instream Flow Incremental Methodology - a Primer for IFIM. Biological Report 29*. National Biological Service, U.S. Department of the Interior, Washington, D.C.

Stanford, J., Lorang, M., and Hauer, F. (2005). The shifting habitat mosaic of river ecosystems. *Verhandlungen der Internationalen Vereinigung für Theoretische und Angewandte Limnologie*, 29:123–136.

Stanford, J. and Ward, J. V. (1993). An ecosystem perspective of alluvial rivers: connectivity and the hyporheic corridor. *Journal of the North American Benthological Society*, 12:48–60.

Statzner, B. and Bêche, L. (2010). Can biological invertebrate traits resolve effects of multiple stressors on running water ecosystems? *Freshwater Biology*, 55:80–119.

Statzner, B., Gore, J. A., and Resh, V. H. (1988). Hydraulic stream ecology: Observed patterns and potential applications. *Journal of the North American Benthological Society*, 7:307–360.

Statzner, B. and Müller, R. (1989). Standard hemispheres as indicators of flow characteristics in lotic benthos research. *Freshwater Biology*, 21:445–459.

Strong, D. (1980). Null hypotheses in ecology. *Synthese*, 43:271–285.

Sundermann, A., Stoll, S., and Haase, P. (2011). River restoration success depends on the species pool of the immediate surroundings. *Ecological applications*, 21:1962–1971.

Talbot, L. (1997). The linkages between ecology and conservation policy. In Pickett, S. T. A., Ostfeld, R. S., Shachak, M., and Likens, G. E., editors, *The Ecological Basis of Conservation: Heterogeneity, Ecosystems, and Biodiversity*. Springer US.

Te Chow, V. (1959). *Open-channel hydraulics*. McGraw-Hill Civil Engineering Series. McGraw-Hill.

Thiery, R. G. (1982). Environmental instability and community diversity*. *Biological Reviews*, 57:691–710.

Thomas, R. and Nicholas, A. (2002). Simulation of braided river flow using a new cellular routing scheme. *Geomorphology*, 43:179 – 195.

Thompson, R. and Townsend, C. (2006). A truce with neutral theory: local deterministic factors, species traits and dispersal limitation together determine patterns of diversity in stream invertebrates. *Journal of Animal Ecology*, 75:476–484.

Tockner, K., Malard, F., and Ward, J. V. (2000). An extension of the Flood Pulse Concept. *Hydrological Processes*, 14:2861–2883.

Tockner, K., Pennetzdorfer, D., Reiner, N., Schiemer, F., and Ward, J. V. (1999). Hydrological connectivity, and the exchange of organic matter and nutrients in a dynamic river–floodplain system (danube, austria). *Freshwater Biology*, 41:521–535.

Tockner, K., Pusch, M., Gessner, J., and Wolter, C. (2011). Domesticated ecosystems and novel communities: challenges for the management of large rivers. *Ecohydrology & Hydrobiology*, 11:167–174.

Toro, E. (2001). *Shock-capturing methods for free-surface shallow flows*. Wiley.

Townsend, C., Dolédec, S., and Scarsbrook, M. (1997). Species traits in relation to temporal and spatial heterogeneity in streams: a test of habitat templet theory. *Freshwater Biology*, 37:367–387.

Townsend, C. R. (1989). The patch dynamics concept of stream community ecology. *Journal of the North American Benthological Society*, 8:36–50.

Townsend, C. R. and Hildrew, A. G. (1976). Field experiments on the drifting, colonization and continuous redistribution of stream benthos. *Journal of Animal Ecology*, 45(3):pp. 759–772.

Turchin, P. (2013). *Complex Population Dynamics : A Theoretical/Empirical Synthesis*. Princeton University Press.

Turner, M., Gardner, R., and O'Neill, R. (2001). *Landscape Ecology in Theory and Practice: Pattern and Process*. Springer.

Underwood, A. (1991). Beyond BACI: experimental designs for detecting human environmental impacts on temporal variations in natural populations. *Marine and Freshwater Research*, 42:569–587.

Underwood, A. (1992). Beyond BACI: the detection of environmental impacts on populations in the real, but variable, world. *Journal of experimental marine biology and ecology*, 161:145–178.

Underwood, A. (1994). On beyond BACI: sampling designs that might reliably detect environmental disturbances. *Ecological applications*, 4:3–15.

Underwood, A. J. (1989). The analysis of stress in natural populations. *Biological Journal of the Linnean Society*, 37:51–78.

Underwood, A. J. (1993). The mechanics of spatially replicated sampling programmes to detect environmental impacts in a variable world. *Australian Journal of Ecology*, 18:99–116.

Underwood, A. J., Chapman, M. G., and Crowe, T. P. (2004). Identifying and understanding ecological preferences for habitat or prey. *Journal of experimental marine biology and ecology*, 300:161–187.

USACE (1993). *River hydraulics. Manual No. 1110-2-1416*. U.S. Army Corps of Engineers.

Vandekerkhove, J. and Cardoso, A. (2010). Alien species and the Water Framework Directive - Questionnaire results. EUR - Scientific and Technical Research series.

Vannote, R. L., Minshall, G. W., Cummins, K. W., Sedell, J. R., and Cushing, C. E. (1980). The river continuum concept. *Canadian Journal of Fisheries and Aquatic Sciences*, 37:130–137.

Vaughan, I. P., Diamond, M., Gurnell, A. M., Hall, K. A., Jenkins, A., Milner, N. J., Naylor, L. A., Sear, D. A., Woodward, G., and Ormerod, S. J. (2009). Integrating ecology with hydromorphology: a priority for river science and management. *Aquatic Conservation: Marine and Freshwater Ecosystems*, 19:113–125.

Verberk, W. C. E. P., van Noordwijk, C. G. E., and Hildrew, A. G. (2013). Delivering on a promise: integrating species traits to transform descriptive community ecology into a predictive science. *Freshwater Science*, 32:531–547.

Vogel, R. M. (2011). Hydromorphology. *Journal of Water Resources Planning and Management*, 137:147–149.

Vosswinkel, N., Mohn, R., Buttschardt, T., Meyer, E., Bünning, I., and Riss, H. (2013). Habitat restoration in an urban channel – the Münstersche Aa – experiences from monitoring of a pilot action. In *Proceedings of the 8th Symposium for European Freshwater Sciences - SEFS 8*.

Ward, J. and Stanford, J. (1983). The serial discontinuity concept of lotic ecosystems. In Fontaine, T. D. and Bartell, S. M., editors, *Dynamic of Lotic Ecosystems*. Ann Arbor Science, Ann Arbor, Michigan.

Ward, J. V. and Stanford, J. A. (1995a). Ecological connectivity in alluvial river ecosystems and its disruption by flow regulation. *Regulated Rivers: Research & Management*, 11:105–119.

Ward, J. V. and Stanford, J. A. (1995b). The serial discontinuity concept: extending the model to floodplain rivers. *Regulated Rivers: Research & Management*, 10:159–168.

Ward, J. V., Tockner, K., Arscott, D. B., and Claret, C. (2002). Riverine landscape diversity. *Freshwater Biology*, 47:517–539.

Wiens, J. (1989). Spatial scaling in ecology. *Functional Ecology*, 3:385–397.

Wiens, J. (2002). Riverine landscapes: taking landscape ecology into the water. *Freshwater Biology*, 47:501–515.

Williams, G. (1974). *Adaptation and natural selection: a critique of some current evolutionary thought*. Princeton University Press: Princeton.

Wittfoth, A. and Zettler, M. (2013). The application of a biopollution index in german baltic estuarine and lagoon waters. *Management of Biological Invasions*, 4:43–50.

Wohl, E., Angermeier, P. L., Bledsoe, B., Kondolf, G. M., MacDonnell, L., Merritt, D. M., Palmer, M. A., Poff, N. L., and Tarboton, D. (2005). River restoration. *Water Resources Research*, 41.

Wolter, C., Lorenz, S., Scheunig, S., Lehmann, N., Schomaker, C., Nastase, A., de Jalón, D. G., Marzin, A., Lorenz, A., Kraková, M., Brabec, K., and Noble, R. (2013). Review on ecological response to hydromorphological degradation and restoration. Deliverable D1.3, Project REFORM, REstoring rivers FOR effective catchment Management. European Commission, 7th Framework Programme, Grant Agreement 282656.

Wyżga, B., Zawiejska, J., Radecki-Pawlik, A., and Hajdukiewicz, H. (2012). Environmental change, hydromorphological reference conditions and the restoration of polish carpathian rivers. *Earth Surface Processes and Landforms*, 37:1213–1226.

Xie, Y. (2013). *animation: a gallery of animations in statistics and utilities to create animations*. R package version 2.2.